Minitab Guide for
Moore and McCabe's

Introduction to the Practice of Statistics

Michael Evans

University of Toronto

W. H. Freeman and Company

New York

Contents

Preface

This Minitab manual is to be used as an accompaniment to *Introduction to the Practice of Statistics,* Fourth Edition, by David S. Moore and George P. McCabe, and to the CD-ROM that accompanies this text. We abbreviate the textbook title as IPS.

Minitab is a statistical software package that was designed especially for the teaching of introductory statistics courses. It is our view that an easy-to-use statistical software package is a vital and significant component of such a course. This permits the student to focus on statistical concepts and thinking rather than computations or the learning of a statistical package. The main aim of any introductory statistics course should always be the "why" of statistics rather than technical details that do little to stimulate the majority of students or, in our opinion, do little to reinforce the key concepts. IPS succeeds admirably in communicating the important basic foundations of statistical thinking, and it is hoped that this manual serves as a useful adjunct to the text.

It is natural to ask why Minitab is advocated for the course. In the author's experience, ease of learning and use are the salient features of the package, with obvious benefits to the student and to the instructor, who can relegate many details to the software. While more sophisticated packages are necessary for higher-level professional work, it is our experience that attempting to teach one of these in a course forces too much attention on technical aspects. The time students need to spend to learn Minitab is relatively small and that it is a great virtue. Further Minitab will serve as a perfectly adequate tool for many of the statistical problems students will encounter in their undergraduate education.

This manual is divided into two parts. Part I is an introduction that provides the necessary details to start using Minitab and in particular how to use worksheets. Not all the material in Part I needs to be absorbed on first reading. We recommend reading I.1–I.10 before starting to use Minitab. The material in I.11 is more for reference and for later reading. References are made to these sections later in the manual and can provide the stimulus to read them. Overall, the introductory Part I also serves as a reference for most of the nonstatistical commands in Minitab.

Part II follows the structure of the textbook. Each chapter is titled and numbered as in IPS. The last two chapters are not in IPS but correspond to optional material included on the CD-ROM. The Minitab commands relevant to doing the problems in each IPS chapter are introduced and their use illustrated. Each chapter concludes with a set of exercises, some of which are modifications of or related to problems in IPS and many of which are new and specifically designed to ensure that the relevant Minitab material has been understood. There are also appendices dealing with some more advanced features of Minitab, such as programming in Minitab and matrix algebra.

Minitab is available in a variety of versions and for different types of computing systems. In writing the manual, we have used Version 13 for Windows, as discussed in the references in Appendix F, but have tried to make the contents of the manual compatible with earlier versions and for versions running under other operating systems. The core of the manual is a discussion of the menu commands while not neglecting to refer to the session commands. Overall, we feel that the manual can be successfully used with most versions of Minitab.

This manual does not attempt a complete coverage of Minitab. Rather, we introduce and discuss those concepts in Minitab that we feel are most relevant for a student studying introductory statistics with IPS. We do introduce some concepts that are, strictly speaking, not necessary for solving the problems in IPS where we feel that they were likely to prove useful in a large number of data analysis problems encountered outside the classroom. While the manual's primary goal is to teach Minitab, generally we want to help develop strong data analytic skills in conjunction with the text and the CD-ROM.

Thanks to Patrick Farace and Chris Spavins of W. H. Freeman and Company for their help and consideration. Also thanks to Rosemary and Heather.

For further information on Minitab software, contact:

Minitab Inc.
3081 Enterprise Drive
State College, PA 16801 USA
ph: 814.328.3280
fax: 814.238.4383
email: Info@minitab.com
URL: http://www.minitab.com

Part I

Minitab for Data Management

New Minitab commands discussed in this part

Calc ▶ Calculator   Calc ▶ Column Statistics
Calc ▶ Make Patterned Data   Calc ▶ Row Statistics
Edit ▶ Copy Cells   Edit ▶ Cut Cells
Edit ▶ Paste Cells   Edit ▶ Select All Cells
Edit ▶ Undo Cut   Edit ▶ Undo Paste
Editor ▶ Enable Command Language Editor ▶ Insert Cells
Editor ▶ Insert Columns   Editor ▶ Insert Rows
Editor ▶ Make Output Editable
File ▶ Exit   File ▶ New
File ▶ Other Files ▶ Export Special Text File ▶ Open Worksheet
File ▶ Other Files ▶ Import Special Text File ▶ Print Session Window
File ▶ Print Worksheet   File ▶ Save Current Worksheet
File ▶ Save Current Worksheet As File ▶ Save Session Window As
Help
Manip ▶ Code   Manip ▶ Concatenate
Manip ▶ Copy Columns   Manip ▶ Display Data
Manip ▶ Erase Variables   Manip ▶ Rank
Manip ▶ Sort   Manip ▶ Stack
Manip ▶ Unstack
Window ▶ Project Manager

1 Manual Overview and Conventions

The manual is divided into two parts. Part I is concerned with getting data into and out of Minitab and giving you the tools necessary to perform various elementary operations on the data so that it is in a form in which you can carry out a statistical analysis. You do not need to understand everything in Part I to begin doing the problems in your course. Part II is concerned with the statistical analysis of the data set and the Minitab commands to do this. The chapters in Part II follow the chapters in *Introduction to the Practice of Statistics*, Fourth Edition, by David S. Moore and George P. McCabe, and to the CD-ROM that accompanies this text (IPS hereafter) and are numbered accordingly. Before you start on Chapter II.1, however, you should read I.1–I.10 and leave I.11 for later reading.

Minitab is a software package that runs on a variety of different types of computers and comes in a number of versions. This manual does not try to describe all the possible implementations or the full extent of the package. We limit our discussion to those features common to the most recent versions of Minitab and, in particular, Versions 12 and 13. Also, we present only those aspects of Minitab relevant to carrying out the statistical analyses discussed in IPS. Of course, this is a fairly wide range of analyses, but the full power of Minitab is not necessary. Depending on the version of Minitab you are using, there may be many more useful features, and we encourage you to learn and use them. Throughout the manual, we point out what some of the additional useful features of Minitab are and how you can go about learning how to use them. Version 13 refers to the most current version of Minitab at the time of writing this manual.

In this manual, special statistical or Minitab concepts will be highlighted in *italic* font. You should be sure that you understand these concepts. We will provide a brief explanation for any terms not defined in IPS. When a reference is made to a Minitab *session command* or *subcommand*, its name will be in **bold** font. Primarily, we will be discussing the *menu commands* that are available in Minitab. Menu commands are accessed by clicking the left button of the mouse on items in lists. We use a special notation for menu commands. For example,

A ▶ B ▶ C

is to be interpreted as left click the command A on the menu bar, then in the list that drops down, left click the command B, and, finally, left click C. The menu commands will be denoted in ordinary font (the actual appearance may vary slightly depending on the version of Windows you use). Any commands that we type and the output obtained will be denoted in `typewriter` font, as will the names of any files used by Minitab, variables, constants, and worksheets.

At the end of each chapter, we provide a few exercises that can be used to make sure you have understood the material. We recommend, however, that whenever possible you use Minitab to do the problems in IPS. While many problems can be done by hand, you will save a considerable amount of time and avoid errors by learning to use Minitab effectively. We also recommend that you try out the Minitab commands as you read about them, as this will ensure full understanding.

2 Accessing and Exiting Minitab

The first thing you should do is find out how to access the Minitab package for your course. This information will come from your instructor, system personnel, or from your software documentation if you have purchased Minitab to run on your own computer.

In some cases, this may mean you type a command such as **minitab** at a computer system prompt and then hit the Enter or Return key on the keyboard after you have logged on, i.e., provided a login name and password to the computer system being used in your course. Typically, you will see the prompt

MTB >

on your screen, and this indicates that you have started a Minitab *session*.

In most cases, you will double click an icon, such as that shown in Display I.1, that corresponds to the Minitab program.

Display I.1: Minitab icon.

Alternatively, you can use the Start button and click on Minitab in the Programs list. In this case, the program opens with a *Minitab window*, such as the one shown in Display I.2. The Minitab window is divided into two sub-windows with the upper window called the *Session window* and the lower one called the *Data window*.

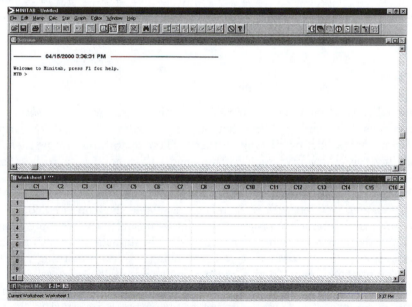

Display I.2: Minitab window.

Left clicking the mouse anywhere on a particular window brings that window to the foreground, i.e., makes it the *active* window, and the border at the top of the window turns dark blue. For example, clicking in the Session window will make the window containing the MTB > prompt active. Alternatively, you can use the command Window ▶ Session in the *menu bar* at the top of the Minitab window to make this window active. You may not see the MTB $>$ prompt in your Session window, and for this manual it is important that you do so. You can ensure that this prompt always appears in your Session window by using Edit ▶ Preferences, doubleclick on Session Window in the Preferences list that comes up, clicking on the Enable radio button under Command Language in the Session Window Preferences, clicking on OK, and clicking on Save. Without the MTB $>$ prompt, you cannot type commands to be executed in the Session window.

In the session window, Minitab *commands* are typed after the MTB $>$ prompt and executed when you hit the Enter or Return key. For example, the first command you should learn is **exit,** as this takes you out of your Minitab session and returns you to the system prompt or operating system. Otherwise, you can access commands using the menu bar (Display I.3) that resides at the top of the Minitab window. For example, you can access the **exit** command using File ▶ Exit. In many circumstances, using the menu commands to do your analyses is easy and convenient, although there are certain circumstances where typing the session commands is necessary. You can also exit by clicking on the × symbol in the upper right-hand corner of the Minitab window. When you exit, you are prompted by Minitab in a dialog window with the question, "Save changes to this Project before closing?" You can safely answer no to this question unless you are in fact using the Projects feature in Minitab as described in Appendix A. In I.8, we will discuss how to save the contents of a Data window before exiting. This is something you will commonly want to do.

File	Edit	Manip	Calc	Stat	Graph	Editor	Window	Help

Display I.3: Menu bar.

Immediately below the menu bar in the Minitab window is the *taskbar*. The taskbar consists of various icons that provide a shortcut method for carrying out various operations by clicking on them. These operations can be identified by holding the cursor over each in turn, and it is a good idea to familiarize yourself with these. Of particular importance are the Cut Cells, Copy Cells, and Paste Cells icons, which are available when a Data window is active. When the operation associated with an icon is not available the icon is faded.

Minitab is an interactive program. By this we mean that you supply Minitab with input data, or tell it where your input data is, and then Minitab responds instantaneously to any commands you give telling it to do something with that data. You are then ready to give another command. It is also possible to run a collection of Minitab commands in a batch program; i.e., several Minitab commands are executed sequentially before the output is returned to the user. The batch version is useful when there is an extensive number of computations to be carried out. You are referred to Appendix C for more discussion of the batch version.

3 Files Used by Minitab

Minitab can accept input from a variety of files and write output to a variety of files. Each file is distinguished by a *file name* and an *extension* that indicates the type of file it is. For example, `marks.mtw` is the name of a file that would be referred to as 'marks' (note the single quotes around the file name) within Minitab. The extension `.mtw` indicates that this is a Minitab worksheet. We describe what a worksheet is in I.5. This file is stored somewhere on the hard drive of a computer as a file called `marks.mtw`.

There are other files that you will want to access from outside Minitab, perhaps to print them out on a printer. Depending on the version of Minitab you are using, to do this, you may have to exit Minitab and give the relevant system print command together with the full path name of the file you wish to print. As various implementations of Minitab differ as to where these files are stored on the hard drive, you will have to determine this information from your instructor or documentation or systems person. For example, in the windows environment the full path name of the file could be

```
c:\Program Files\MTBWIN\Data\marks.mtw
```

or something similar. This path name indicates that the file `marks.mtw` is stored on the C hard drive in the directory called `Program Files\Mtbwin\Data`. We will discuss several different types of files in this chapter.

In many versions of Minitab, there are restrictions on file names. For example, in earlier versions a file name can be at most eight characters in length using any symbols except # and ' and the first character cannot be a blank. There is no length restriction on file names in Versions 12 or 13. It is generally best to name your files so that the file name reflects its contents. For example, the file name `marks` may refer to a data set composed of student marks in a number of courses.

4 Getting Help

At times, you may want more information about a command or some other aspect of Minitab than this manual provides, or you may wish to remind yourself of some detail that you have partially forgotten. Minitab contains an online manual that is very convenient. You can access this information directly by clicking on Help in the Menu bar and using the table of _Contents or doing a Search of the manual for a particular concept.

From the MTB > prompt, you can use the **help** command for this purpose. Typing **help** followed by the name of the command of interest and hitting Enter will cause Minitab to produce relevant output. For example, asking for help on the command **help** itself via the command

```
MTB >help help
```

will give you an overview of what help information can be accessed on your system. The **help** command should be used to find out about session commands.

5 The Worksheet

The basic structural component of Minitab is the *worksheet*. Basically, the worksheet can be thought of as a big rectangular array, or matrix, of *cells* organized into rows and columns as in the Data window of Display I.2. Each cell holds one piece of data. This piece of data could be a number, i.e. *numeric data*, or it could be a sequence of characters, such as a word or an arbitrary sequence of letters and numbers, i.e., *text data*. Data often comes as numbers, such as 1.7, 2.3, ... but sometimes it comes in the form of a sequence of characters, such as black, brown, red, etc. Typically, sequences of characters are used as identifiers in classifications for some variable of interest, e.g., color, gender. A piece of text data can be up to 80 characters in length in Minitab. Version 13 also allows for *date data*, which is data especially formatted to indicate a date, for example, 3/4/97. We will not discuss date data.

If possible, try to avoid using text data with Minitab, i.e., make sure all the values of a variable are numbers, as dealing with text data in Minitab is more difficult. For example, denote colors by numbers rather than by names. Still there will be applications where data comes to you as text data, e.g., in a computer file, and it is too extensive to convert to numeric data. So we will discuss how to input text data into a Minitab worksheet, but we recommend that in such cases you convert this to numeric data, using the methods of I.11.3, once it has been input. In Version 13 of Minitab it is somewhat easier to deal with text data than earlier versions, and this proviso is not as necessary.

Display I.4 provides an example of a worksheet. Notice that the columns are labeled C1, C2, etc. and the rows are labeled 1, 2, 3, etc. We will refer to the worksheet depicted in Display I.4 as the **marks** worksheet hereafter and will use it throughout Part I to illustrate various Minitab commands and operations.

Data arises from the process of taking measurements of variables in some real-world context. For example, in a population of students, suppose that we are conducting a study of academic performance in a Statistics course. Specifically, suppose that we want to examine the relationship between grades in Statistics, grades in a Calculus course, grades in a Physics course and gender. So we collect the following information for each student in the study: student number, grade in Statistics, grade in Calculus, grade in Physics, and gender. Therefore, we have 5 variables – student number and the grades in the three subjects are *numeric variables*, and gender is a *text variable*. Let us further suppose that there are 10 students in the study.

Display I.4 gives a possible outcome from collecting the data in such a study. Column C1 contains the student number (note that this is a categorical variable even though it is a number). The student number primarily serves as an identifier so that we can check that the data has been entered correctly. This is something you should always do as a first step in your analysis. Columns C2–C4 contain the student grades in their Statistics, Calculus, and Physics courses and column C5 contains the gender data. Notice that a column contains the values collected for a single variable, and a row contains the values of all the variables for a single student. Sometimes, a row is referred to as an *observation* or *case*. Observe that the data for this study occupies a 10×5 subtable of the full worksheet. All of the other blank entries of the worksheet can be ignored, as they are undefined.

Display I.4: The marks worksheet.

There will be limitations on the number of columns and rows you can have in your worksheet, and this depends on the particular implementation of Minitab you are using. So if you plan to use Minitab for a large problem, you should check with the system person or further documentation to see what these are. For example, in some versions of Minitab there is a limitation of 5000 cells. So there can be one variable with 5000 values in it, or 50 variables with 100 values each, etc.

Associated with a worksheet is a table of *constants*. Typically, these are numbers that you want to use in some arithmetical operation applied to every value in a column. For example, you may have recorded heights of people in inches and want to convert these to heights in centimeters. You must multiply every height by the value 2.54. The Minitab constants are labeled K1, K2, etc. Again, there are limitations on the number of constants you can associate with a worksheet. For example, in many versions there can be at most 1000 constants. So to continue with the above problem, we might assign the value 2.54 to K1. In I.7.4, we show how to make such an assignment, and in I.10.1 we show how to multiply every entry in a column by this value.

In Version 13 of Minitab, there is an additional structure beyond the worksheet called the *project*. A project can have multiple worksheets associated with it. Also, a project can have associated with it various graphs and records of the commands you have typed and the output obtained while working on the worksheets. Projects, which are discussed in Appendix A, can be saved and retrieved for later work. Projects .

6 Minitab Commands

We will now begin to introduce various Minitab commands to get data into a worksheet, edit a worksheet, perform various operations on the elements of a worksheet, and save and access a saved worksheet. Before we do, however, it is useful to know something about the basic structure of all Minitab commands. Associated with every command is of course its *name*, as in File ▶ Exit and Help. Most commands also take *arguments*, and these arguments are column names, constants, and sometimes file names.

Commands can be accessed by making use of the File, Edit, Manip, Calc, Stat, Graph and Editor entries in the menu bar. Clicking any of these brings up a list of commands that you can use to operate on your worksheet. The lists that appear may depend on which window is active, e.g., either a Data window or the Session window. Unless otherwise specified, we will always assume that the Session window is active when discussing menu commands. If a command name in a list is faded, then it is not available.

Typically, using a command from the menu bar requires the use of a *dialog box* or *dialog window* that opens when you click on a command in the list. These are used to provide the arguments and subcommands to the command and specify where the output is to go. Dialog boxes have various boxes that must be filled in to correctly execute a command. Clicking in a box that needs to be filled in typically causes a *variable list* to appear in the left-most box, of all items in the active worksheet that can be placed in that box. Double clicking on items in the variable list places them in the box, or, alternatively, you can type them in directly. When you have filled in the dialog box and clicked OK, the command is printed in the Session window and executed. Any output is also printed in the Session window. Dialog boxes have a Help button that can be used to learn how to make the entries.

For example, suppose that we want to calculate the *mean* of column C2 in the worksheet marks. Then the command Calc ▶ Column Statistics brings up the dialog box shown in Display I.5. Notice that the radio button Sum is filled in. Clicking the radio button labelled Mean results in this button being filled in and the Sum button becoming empty. Whichever button is filled in will result in that statistic being calculated for the relevant columns when we finally implement the command by clicking OK.

Currently, there are no columns selected, but clicking in the Input variable box brings up a list of possible columns in the display window on the left. The results of these operations are shown in Display I.6. We double click on C2 in the variable list, which places this entry in the Input variable box as shown in Display I.7. Alternatively, we could have simply typed this entry into the box. After clicking the OK button, we obtain the output

```
Mean of C2 = 69.900
```

in the Session window.

Display I.5: Initial view of the dialog box for Column Statistics.

Display I.6: View of the dialog box for Column Statistics after selecting Mean and bringing up the variable list.

Column Statistics

C1	**Statistic**	
C2	○ S<u>u</u>m	○ M<u>e</u>dian
C3	◉ <u>M</u>ean	○ Sum of s<u>q</u>uares
C4	○ <u>S</u>tandard deviation	○ N <u>t</u>otal
	○ M<u>i</u>nimum	○ <u>N</u> nonmissing
	○ Ma<u>x</u>imum	○ N missing
	○ <u>R</u>ange	

Input <u>v</u>ariable: | C2 |

Store resu<u>l</u>t in: | | [Optional]

[Select]

[Help]   [<u>O</u>K]   [Cancel]

Display I.7: Final view of the dialog box for Column Statistics.

Quite often, it is faster and more convenient to simply type your commands directly into the Session window. Sometimes, it is necessary to use the Session window approach, but for many commands the menu bar is available. So we now describe the use of commands in the Session window.

The basic structure of such a command with n arguments is

command name $E_1, E_2, ..., E_n$

where E_i is the *ith* argument. Alternatively, we can write

command name $E_1 \ E_2 \ ... \ E_n$

if we don't want to type commas. Conveniently, if the arguments $E_1, E_2, ..., E_n$ are consecutive columns in the worksheet, we have the following short-form

command name $E_1 - E_n$

which saves even more typing and accordingly decreases our chance of making a typing mistake. If you are going to type a long list of arguments and you don't want them all on the same line, then you can type the *continuation symbol* & where you want to break the line and then hit Enter. Minitab responds with the prompt

CONT>

and you continue to type argument names. The command is executed when you hit Enter after an argument name without a continuation character following it.

Many commands can, in addition, be supplied with various subcommands that alter the behavior of the command. The structure for commands with subcommands is

command name $E_1 \ldots E_{n_1}$;
subcommand name $E_{n_1+1} \ldots E_{n_2}$;
$$\vdots$$
subcommand name $E_{n_{k-1}+1} \ldots E_{n_k}$.

Notice that when there are subcommands each line ends with a semicolon until the last subcommand, which ends with a period. Also, subcommands may have arguments. When Minitab encounters a line ending in a semicolon it expects a subcommand on the next line and changes the prompt to

`SUBC >`

until it encounters a period, whereupon it executes the command. If while typing in one of your subcommands you suddenly decide that you would rather not execute the subcommand — perhaps you realize something was wrong on a previous line — then type **abort** after the `SUBC >` prompt and hit Enter. As a further convenience, it is worth noting that you need to only type in the first four letters of any Minitab command or subcommand.

For example, to calculate the mean of column C2 in the worksheet marks we can use the **mean** command in the Session window, as in

`MTB > mean c2`

and we obtain the same output in the Session window as before.

There are two additional ways in which you can input commands to Minitab. Instead of typing the commands directly into the Session window, you can also type these directly into the Command Line Editor, which is available via Edit ▶ Command Line Editor. Multiple commands can then be typed directly into a box that pops up and executed when the Submit Commands button is clicked. Output appears in the Session window. Also, many commands are available on a *toolbar* that lies just below the menu bar at the top of the Minitab window. There is a different toolbar depending upon which window is active. We give a brief discussion of some of the features available in the toolbar in later sections.

7 Entering Data into a Worksheet

There are various methods for entering data into a worksheet. The simplest approach is to use the *Data window* to enter data directly into the worksheet by clicking your mouse in a cell and then typing the corresponding data entry and hitting Enter. Remember that you can make a Data window active by clicking anywhere in the window or by using Windows in the menu bar. If you type any character that is not a number, Minitab automatically identifies the column containing that cell as a text variable and indicates that by appending T to the column name, e.g., C5-T in Display I.4. You do not need to append the T when referring to the column. Also, there is a *data direction arrow* in the upper left corner of the data window that indicates the direction the cursor moves after you hit Enter. Clicking on it alternates between row-wise and column-wise data entry. Certainly, this is an easy way to enter data when it is suitable. Remember, columns are variables and rows are observations! Also, you can have multiple data windows open and move data between them. Use the command File ▶ New to open a new worksheet.

7.1 Importing Data

If your data is in an external file (not an .mtw file), you will need to use File ▶ Other Files ▶ Import Special Text to get the data into your worksheet. For example, suppose in the file marks.txt we have the following data recorded, just as it appears.

```
12389 81 85 78
97658 75 72 62
53546 77 83 81
55542 63 42 55
11223 71 82 67
77788 87 56 *
44567 23 45 35
32156 67 72 81
33456 81 77 88
67945 74 91 92
```

Each row corresponds to an observation, with the student number being the first entry, followed by the marks in the student's Statistics, Calculus, and Physics courses. These entries are separated by blanks.

Notice the * in the sixth row of this data file. In Minitab, a * signifies a *missing numeric value*, i.e., a data value that for some reason is not available. Alternatively, we could have just left this entry blank. A *missing text value* is simply denoted by a blank. Special attention should be paid to missing values. In general, Minitab statistical analyses ignore any cases that contain missing data except that the output of the command will tell you how many cases were ignored because of missing data. It is important to pay attention to this information. If your data is riddled with a large number of missing values, your analysis may be based on very few observations — even if you have a large data set!

When data in such a file is *blank-delimited* like this it is very easy to read in. After the command File ▶ Other Files ▶ Import Special Text, we see the dialog box shown in Display I.8 minus C1–C4 in the Store data in column(s): box. We typed C1-C4 into this window to indicate that we want the data read in to be stored in these columns. Note that it doesn't matter if we use lower or upper case for the column names, as Minitab is not case sensitive. After clicking OK, we see the dialog box depicted in Display I.9, which we use to indicate from which file we want to read the data. Note that if your data is in .txt files rather than .dat files, you will have to indicate that you want to see these in the Files of type box by selecting Text Files or perhaps All Files. Clicking on `marks.txt` results in the data being read into the worksheet.

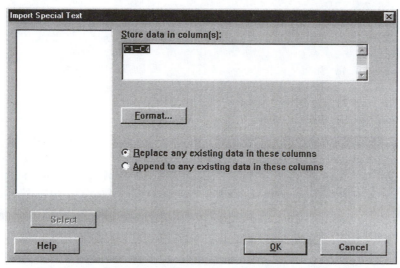

Display I.8: Dialog box for importing data from external file.

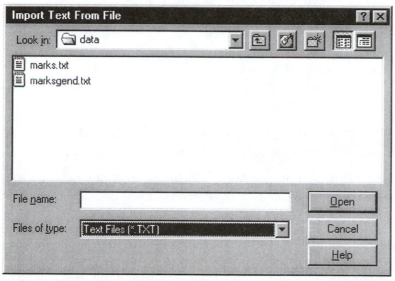

Display I.9: Dialog box for selecting file from which data is to be read in.

Of course, this data set does not contain the text variable denoting the student's gender. Suppose that the file `marksgend.txt` contains the following data exactly as typed.

```
12389 81 85 78 m
97658 75 72 62 m
53546 77 83 81 f
55542 63 42 55 m
11223 71 82 67 f
77788 87 56  * f
44567 23 45 35 m
32156 67 72 81 m
33456 81 77 88 f
67945 74 91 92 f
```

As this file contains text data in the fifth column, we must tell Minitab how the data is *formatted* in the file. To access this feature we click on the Format button in the dialog box shown in Display I.8. This brings up the dialog box shown in Display I.10.

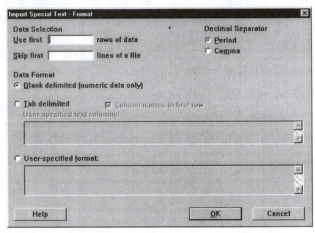

Display I.10: Initial dialog box for formatted input.

To indicate that we will specify the format, we click the radio button User-specified format and fill the particular format into the box as shown in Display I.11. The format statement says that we are going to read in the data according to the following rule: a numeric variable occupying 5 spaces and with no decimals, followed by a space, a numeric variable occupying 2 spaces with no decimals, a space, a numeric variable occupying 2 spaces with no decimals, a space, a numeric variable occupying 2 spaces with no decimals, a space, and a text variable occupying 1 space. This rule must be rigorously adhered to or errors will occur. So the rules you need to remember if you use formatted input are that ak indicates a text variable occupying k spaces, kx indicates k spaces, and fk.l indicates a numeric variable occupying k spaces, of which l are to the right of the decimal point. Note if a data value does not fill up the full number of spaces allotted to it in the format statement, it must be right justified in its field. Also, if a decimal point is included in the number, this occupies one of the spaces allocated to the variable and similarly for a negative or plus sign. There are many other features to formatted input that we will not discuss here. Use the Help button in the dialog box for information on these features. Finally, clicking on the OK button reads this data into a worksheet as depicted in Display I.4. Typically, we try to avoid the use of formatted input because it is somewhat cumbersome, but sometimes we must use it.

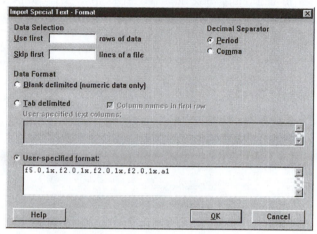

Display I.11: Dialog box for formatted input with the format filled in.

In the session environment, the **read** command is available for inputting data into a worksheet with capabilities similar to what we have described. For example, the commands

```
MTB >read c1-c4
DATA>12389 81 85 78
DATA>97658 75 72 62
DATA>53546 77 83 81
DATA>55542 63 42 55
DATA>11223 71 82 67
DATA>77788 87 56 *
DATA>44567 23 45 35
DATA>32156 67 72 81
DATA>33456 81 77 88
DATA>67945 74 91 92
DATA>end
 10 rows read.
```

place the first four columns into the marks worksheet. After typing read c1-c4 after the MTB > prompt and hitting Enter, Minitab responds with the DATA> prompt, and we type each row of the worksheet in as shown. To indicate that there is no more data, we type **end** and hit Enter. Similarly, we can enter text data in this way but can't combine the two unless we use a **format** subcommand. We refer the reader to **help** for more description of how this command works.

7.2 Patterned Data

Often, we want to input *patterned data* into a worksheet. By this we mean that the values of a variable follow some determined rule. We use the command Çalc ▶ Make Patterned Data for this. For example, implementing this command with the entries in the dialog box depicted in Display I.12 adds a column C6 to the marks worksheet where the sequence $0, 0.5, 1.0, 1.5, 2.0$ is repeated twice. For this we entered 0 in the From first value box, a 2 in the To last value box, a .5 in the In steps of box, a 1 in the List each value box, and a 2 in the List the whole sequence box. Basically, we can start a sequence at any number m and successively increment this with any number $d > 0$ until the next addition would exceed the last value n prescribed, repeat each element l times, and finally repeat the whole sequence k times.

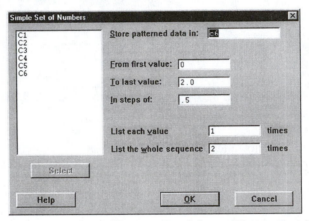

Display I.12: Dialog box for making patterned data with some entries filled in.

There is some shorthand associated with patterned data that can be very convenient. For example, typing $m : n$ in a Minitab command is equivalent to typing the values $m, m + 1, \ldots, n$ when $m < n$ and $m, m - 1, \ldots, n$ when $m > n$ and m when $m = n$. The expression $m : n/d$, where $d > 0$, expands to a list as above but with the increment of d or $-d$, whichever is relevant, replacing 1 or -1. If $m < n$ then d is added to m until the next addition would exceed n and if $m > n$ then d is subtracted from m until the next subtraction would be lower than n. The expression $k(m : n/d)$ repeats $m : n/d$ for k times while $(m : n/d)l$ repeats each element in $m : n/d$ for l times. The expression $k(m : n/d)l$ repeats $(m : n/d)l$ for k times.

The **set** command is available in the session window to input patterned data. For example, suppose we want C6 to contain the 10 entries 1, 2, 3, 4, 5, 5, 4, 3, 2, 1. The command

```
MTB >set c6
DATA>1:5
DATA>5:1
DATA>end
```

does this. Also, we can add elements in parentheses. For example, the command

```
MTB >set c6
DATA>(1:2/.5 4:3/.2)
DATA>end
```

creates the column with entries 1.0, 1.5, 2.0, 4.0, 3.8, 3.6, 3.4, 3.2, 3.0. The multiplicative factors k and l can also be used in such a context. Obviously, there is a great deal of scope for entering patterned data with **set**. The general syntax of the set command is

set E_1

where E_1 is a column.

7.3 Printing Data in the Session Window

Once we have entered the data into the worksheet, we should always check that we have made the entries correctly. Typically, this means printing out the worksheet and checking the entries. The command Manip ▶ Display Data will print the data you ask for in the Session window. For example, with the worksheet **marks** the dialog box pictured in Display I.13 causes the contents of this worksheet to be printed when we click on OK. We selected which variables to print by first clicking in the Columns, constants, and matrices to display box and then double clicking on the variables in the variable list on the left.

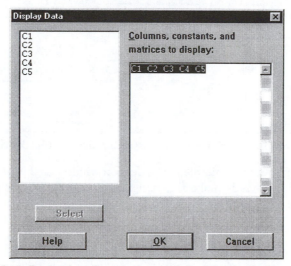

Display I.13: Dialog box for printing worksheet in the Session window.

The **print** command is available in the Session window and is often convenient to use. The general syntax for the **print** command is

print $E_1 \ldots E_m$

where $E_1, \ldots, E_m$ are columns and constants.

7.4 Assigning Constants

To enter constants, we use the Calc ▶ Calculator command and fill in the dialog box appropriately. For example, suppose we want to assign the values k1=.5, k2=.25 and k3=.25 to the constants k1, k2, and k3. These could serve as weights to calculate a weighted average of the marks in the marks worksheet. Then the Calc ▶ Calculator command leads to the dialog box displayed in Display I.14, where we have typed k1 into the Store result in variable box and the value .5 into the Expression box. Clicking on OK then makes the assignment. Note that we can assign text values to constants by enclosing the text in double quotes. We will talk about further features of Calculator later in this manual. Similarly, we assign values to k2 and k3.

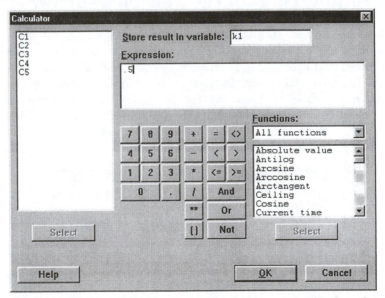

Display I.14: Filled in dialog box for assigning the constant k1 the value .5.

The **let** command is available in the Session window and is quite convenient. The following commands make this assignment and then we check, using the **print** command, that we have entered the constants correctly.

```
MTB >let k1=.5
MTB >let k2=.25
MTB >let k3=.25
MTB >print k1-k3
K1  0.500000
K2  0.250000
K3  0.250000
```

Also, we can assign constants text values. For example,

```
MTB >let k4=''result''
```

assigns K4 the value `result`. Note the use of double quotes.

7.5 Naming Variables and Constants

It often makes sense to give the columns and constants names rather than just referring to them as C1, C2, ..., K1, K2, etc. This is especially true when there are many variables and constants, as it would be easy to slip and use the wrong column in an analysis and then wind up making a mistake. To assign a name to a variable simply go to the blank cell at the top of the column in the worksheet corresponding to the variable and type in an appropriate name. For example, we have used `studid`, `statistics`, `calculus`, `physics`, and `gender` for the names of C1, C2, C3, C4, and C5, respectively, and these names appear in Display I.15.

Display I.15: Worksheet marks with named variables.

In the Session window, the **name** command is available for naming variables and constants. For example, the commands

```
MTB >name c1 'studid' c2 'stats' c3 'calculus' &
CONT>c4 'physics' c5 'gender' &
CONT>k1 'weight1' k2 'weight2' k3 'weight3'
```

give the names studid to C1, stats to C2, calculus to C3, physics to C4, gender to C5, weight1 to K1, weight2 to K2, and weight3 to K3. Notice that we have made use of the continuation character & for convenience in typing in the full input to **name**. When using the variables as arguments just enclose the names in single quotes. For example,

```
MTB >print 'studid' 'calculus'
```

prints out the contents of these variables in the Session window.

Variable and constant names can be at most 31 characters in length, cannot include the characters # and ' and cannot start with a leading blank or *. Recall that Minitab is not case sensitive, so it does not matter if we use lower or upper case letters when specifying the names.

7.6 Information about a Worksheet

We can get information on the data we have entered into the worksheet by using the **info** command in the Session window. For example, we get the following results based on what we have entered into the **marks** worksheet so far.

```
MTB >info
    Column    Name      Count Missing
A  C1        studid        10       0
   C2        stats         10       0
   C3        calculus      10       0
   C4        physics       10       1
A  C5        gender        10       0
   Constant  Name      Value
   K1        weight1   0.500000
   K2        weight2   0.250000
   K3        weight3   0.250000
```

Notice that the **info** command tells us how many missing values there are and in what columns they occur and also the values of the constants.

This information can also be accessed directly from the *Project Manager window* via Window ▶ Project Manager.

7.7 Editing a Worksheet

It often happens that after data entry we notice that we have made some mistakes or we obtain some additional information, such as more observations. So far, the only way we could change any entries in the worksheet or add some rows is to reenter the whole worksheet!

Editing the worksheet is straightforward because we simply change any cells by retyping their entries and hitting the Enter key. We can add rows and columns at the end of the worksheet by simply typing new data entries in the relevant cells. To insert a row before a particular row, simply click on any entry in that row and then the menu command Editor ▶ Insert Rows. Fill in the blank entries in the new row. To insert a column before a particular column, simply click on any entry in that column and then the menu command Editor ▶ Insert Columns. Fill in the blank entries in the new column. To insert a cell before a particular cell, simply click on any entry in that cell and the menu command Editor ▶ Insert Cells. Fill in the blank entry in the new cell that appears in place of the original with all other cells in that column — and only that column — pushed down.

If you wish to clear a number of cells in a block, click in the cell at the start of the block, and holding the mouse key down, drag the cursor through the block so that it is highlighted in black. Click on the Cut Cells icon on the Minitab *taskbar,* and all the entries will be deleted. Cells immediately below the block move up to fill in the vacated places. A convenient method for clearing all the data entries in a worksheet, with the relevant Data window active, is to use the command Edit ▶ Select All Cells, which causes all the cells to be highlighted, and click on the Cut Cells icon. Always save the contents of the current worksheet before doing this unless you are absolutely sure you don't need the data again. We discuss how to save the contents of a worksheet in I.8.1.

To copy a block of cells, click in the cell at the start of the block and, holding the mouse key down, drag the cursor through the block so that it is highlighted in black, but, instead of hitting the backspace key, use the command Edit ▶ Copy Cells or click on the Copy Cells icon on the Minitab taskbar. The block of cells is now copied to your clipboard. If you not only want to copy a block of cells to your clipboard but remove them from the worksheet, use the command Edit ▶ Cut Cells or the Cut Cells icon on the Minitab taskbar instead. Note that any cells below the removed block will move up to replace these entries. To paste the block of cells into the worksheet, click on the cell before which you want the block to appear or that is at the start of the block of cells you wish to replace and issue the command Edit ▶ Paste Cells, or use the Paste Cells icon on the Minitab taskbar. A dialog box appears as in Display I.16, where you are prompted as to what you want to do with the copied block of cells. If you feel that a cutting or pasting was in error, you can undo this operation by using Edit ▶ Undo Cut or Edit ▶ Undo Paste, respectively, or use the Undo icon on the Minitab taskbar.

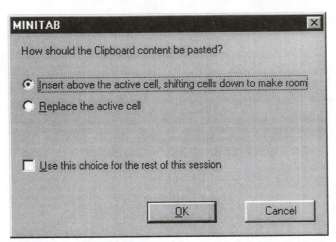

Display I.16: Dialog box that determines how a block of copied cells is used, whether being inserted into a worksheet or replacing a block of cell of the same size.

An alternative approach is available for copying operations using Manip ▶ Copy Columns and filling in the dialog box appropriately. For example, suppose we want to copy all the entries in the marks worksheet in rows 5 and 8 of columns C2 and C4 and place these in columns C7 and C8. The dialog box shown in Display I.17 would result in all the entries in columns C2 and C4 being copied to C7 and C8. To prevent this, we click on the Use Rows button, which brings up the dialog box shown in Display I.18. Clicking on the Use rows radio button and filling in the associated box with the entries 5 and 8 specifies that only entries in the fifth and eighth rows will be copied. Clicking on the OK buttons in these dialog boxes then completes the operation.

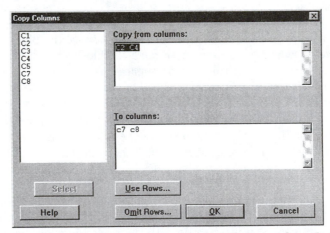

Display I.17: Dialog box for copying entries in columns and pasting them.

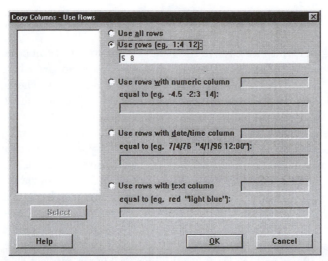

Display I.18: Dialog box to select rows from columns to be copied.

One can also delete selected rows from specified columns using Manip ▶ Delete Rows and filling in the dialog box appropriately. Notice, however, that whenever we delete a cell, the contents of the cells beneath the deleted one in that column simply move up to fill the cell. The cell entry does not become missing; rather, cells at the bottom of the column become undefined! If you delete an entire row, this is not a problem because the rows below just shift up. For example, if we delete the third row then in the new worksheet, after the deletion, the third row is now occupied by what was formerly the fourth row. Therefore, you should be very careful, when you are not deleting whole rows, to ensure that you get the result you intended.

Note that if you should delete all the entries from a column, this variable is still in the worksheet, but it is empty now. If you wish to delete a variable and all its entries, this can be accomplished from Manip ▶ Erase Variables and filling in the dialog box appropriately. This is a good idea if you have a lot of variables and no longer need some of them.

There are various commands in the Session window available for carrying out these editing operations. For example, the **restart** command in the Session window can be used to remove all entries from a worksheet. The **let** command allows you to replace individual entries. For example,

```
MTB > let c2(2)=3
```

assigns the value 3 to the second entry in the column C2. The **copy** command can be used to copy a block of cell from one place to another. The **insert** command allows you to insert rows or observations anywhere in the worksheet. The **delete** command allows you to delete rows. The **erase** command is available for the deletion of columns or variables from the worksheet. As it is more convenient to edit a worksheet by directly working on the worksheet and using the menu commands, we do not discuss these commands further here.

8 Saving, Retrieving, and Printing

Quite often, you will want to save the results of all your work in creating a worksheet. If you exit Minitab before you save your work, you will have to reenter everything. So we recommend that you always save. To use the commands of this section make sure that the Worksheet window of the worksheet in question is active.

Use File ▶ Save Current Worksheet to save the worksheet with its current name, or the default name if it doesn't have one. If you want to provide a name or store the worksheet in a new location, then use File ▶ Save Current Worksheet As and fill in the dialog box depicted in Display I.19 appropriately. The Save in box at the top contains the name of the folder in which the worksheet will be saved once you click on the Save button. Here the folder is called **data**, and you can navigate to a new folder using the Up One Level button immediately to the right of this box. The next button takes you to the Desktop and the third button allows you to create a subfolder within the current folder. The box immediately below contains a list of all files of type .mtw in the current folder. You can select the type of file to display by clicking on the arrow in the Save as type box, which we have done here, and click on the type of file you want to display that appears in the drop-down list. There are several possibilities including saving the worksheet in other formats, such as Excel. Currently, there is only one .mtw file in the folder **data** and it is called **marks.mtw**. If you want to save the worksheet with a different name, type this name in the File name box and click on the Save button.

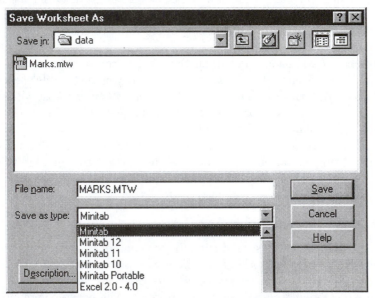

Display I.19: Dialog box for saving a worksheet.

To retrieve a worksheet, use File ▶ Open Worksheet and fill in the dialog box as depicted in Display I.20 appropriately. The various windows and buttons in this dialog box work as described for the File ▶ Save Current Worksheet As command, with the exception that we now type the name of the file we want to open in the File name box and click on the Open button.

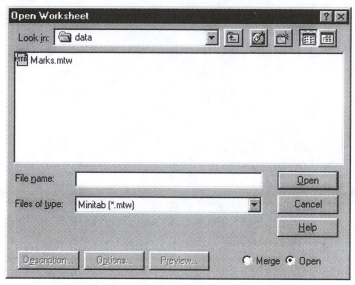

Display I.20: Dialog box for retrieving a worksheet.

To print a worksheet, use the command File ▶ Print Worksheet. The dialog box that subsequently pops up allows you to control the output in a number of ways.

It may be that you would prefer to write out the contents of a worksheet to an external file that can be edited by an editor or perhaps used by some other program. This will not be the case if we save the worksheet as an .mtw file as only Minitab can read these. To do this, use the command File ▶ Other Files ▶ Export Special Text, filling in the dialog box and specifying the destination file when prompted. For example, if we want to save the contents of the marks worksheet, this command results in the dialog box of Display I.21 appearing. We have entered all five columns into the Columns to export box and have not specified a format so the columns will be stored in the file with single blanks separating the columns. Clicking the OK button results in the dialog box of Display I.22 appearing. Here, we have typed in the name marks.dat to hold the contents. Note that while we have chosen a .dat type file, we also could have chosen a .txt type file. Clicking on the Save button results in a file marks.dat being created in the folder data with contents as displayed in Display I.23.

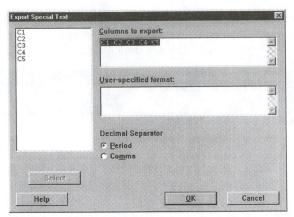

Display I.21: Dialog box for saving the contents of a worksheet to an external (non-Minitab) file.

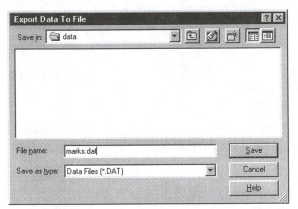

Display I.22: Dialog box for selecting external file to hold contents of a worksheet.

```
12389    81    85    78    m
97658    75    72    62    m
53546    77    83    81    f
55542    63    42    55    m
11223    71    82    67    f
77788    87    56     *    f
44567    23    45    35    m
32156    67    72    81    m
33456    81    77    88    f
67945    74    91    92    f
```

Display I.23: Contents of the file marks.dat.

In the Session window, the commands **save** and **retrieve** are available for saving and retrieving a worksheet in the .mtw format and the command **write** is available for saving a worksheet in an external file. We refer the reader to **help** for a description of how these commands work.

9 Recording and Printing Sessions

Sometimes, it is useful — e.g., when you have to hand in an assignment — to maintain a record of all the commands you used, the output you obtained, and any comments you want to make on what you are doing in a Minitab session. Note that after executing a menu command the relevant Session window commands are automatically typed in the Session window.

To use the commands for saving or printing the Session window first make sure that the Session window is active. If you issue the menu command Editor ▶ Output Editable first, you can edit the Session window contents before saving or printing its contents simply by typing or erasing text in the Session window. You can turn this feature off using the same command. To save the contents of a Session window use File ▶ Save Session Window As and fill in the dialog box appropriately. Note that the saved file is in the .txt format unless you make a different choice in the Save as type box. To print the contents of the Session window use File ▶ Print Session Window.

In the Session window, the **outfile** command is available for recording the full or partial contents of a Minitab session. We refer the reader to **help** for a description of how this command works.

10 Mathematical Operations

When carrying out a data analysis a statistician is often called upon to transform the data in some way. This may involve applying some simple transformation to a variable to create a new variable — e.g., take the natural logarithm of every grade in the **marks** worksheet — to combining several variables together to form a new variable — e.g., calculate the average grade for each student in the **marks** worksheet. In this section, we present some of the ways of doing this.

10.1 Arithmetical Operations

Simple arithmetic can be carried out on the columns of a worksheet using the arithmetical operations of addition +, subtraction −, multiplication *, division /, and exponentiation ** via the Calc ▶ Calculator command. When columns are added together, subtracted one from the other, multiplied together, divided one by the other (make sure there are no zeros in the denominator column), or one column exponentiates another, these operations are always performed component-wise. For example, C1*C2 means that the *ith* entry of C1 is multiplied by the *ith* entry of C2; etc. Also, make sure that the columns on which you are going to perform these operations correspond to numeric variables! While these operations have the order of precedence **, */, +−, parentheses ( ) can and should be used to ensure an unambiguous result. For example, suppose in the **marks** worksheet we want to create a new variable by taking the average of the Statistics and Calculus grades and then subtracting this from the Physics grade and placing the result in C6. Filling in the dialog box, corresponding to Ç̲alc ▶ Ca̲lculator, as shown in Display I.24 accomplishes this when we click on the O̲K button.

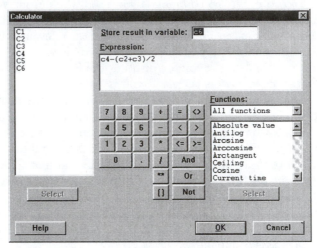

Display I.24: Dialog box for carrying out mathematical calculations.

Note that we can either type the relevant expression into the E̲xpression box or use the buttons and double clicking on the relevant columns. Further, we type the column where we wish to store the results of our calculation in the S̲tore result in variable box. These operations are done on the corresponding entries in each column; corresponding entries in the columns are operated on according to the formula we have specified, and a new column of the same length containing all the outcomes is created. Note that the sixth entry in C6 will be * — missing — because this entry was missing for C4.

These kinds of operations can also be carried out directly in the Session window using the **let** command, and in some ways this is a simpler approach. For example, the session command

```
MTB >let c6=c4-(c2+c3)/2
```

accomplishes this.

We can also use these arithmetical operations on the constants K1, K2, etc., and numbers to create new constants or use the constants as *scalars* in operations with columns. For example, suppose that we want to compute the weighted average of the Statistics, Calculus, and Physics grades where Statistics gets twice the weight of the other grades. Recall that we created, as part of the `marks` worksheet, the constants `weight1` = .5, `weight2` = .25, and `weight3` = .25 in K1, K2, and K3, respectively. So this weighted average is computed via the command

```
MTB >let c7='weight1'*'stats'+'weight2'*'calculus'&
CONT>+'weight3'*'physics'
```

and the result is placed in C7. We have used the continuation character & for convenience in this computation. Alternatively, we could have used the Calc ▶ Calculator command as above for this.

10.2 Mathematical Functions

Various mathematical functions are available in Minitab. For example, suppose we want to compute the natural logarithm of the Statistics mark for each student. Using the Calc ▶ Calculator command with the dialog box as in Display I.25 accomplishes this.

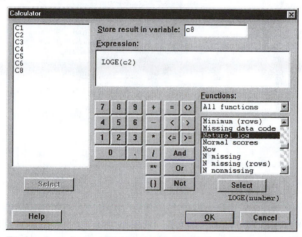

Display I.25: Dialog box for mathematical calculations illustrating the use of the natural logarithm function.

A complete list of such functions is given in the Functions window when All functions is in the window directly above the list.

The same result can be obtained using the session command **let** and the natural logarithm function **loge**. For example,

```
MTB >let c8=loge(c2)
```

calculates the natural log of every entry in c2 and places the results in C8. There are a number of such functions and a complete list is provided in Appendix B.1. These functions can be applied to numbers as well as constants. If you want to know the sine of the number 3.4, then

```
MTB >let k4=sin(3.4)
MTB >print k4
K4 -0.255541
```

gives the value.

10.3 Column and Row Statistics

There are various *column statistics* that compute a single number from a column
by operating on all of the elements in a column. For example, suppose that we
want the mean of all the Statistics marks, i.e., the mean of all the entries in C2.
The command Çalc ▶ Çolumn Statistics produces the dialog box of Display I.26
where we have selected Mean as the particular statistic to compute and C2 as
the column to use. Clicking QK causes the mean of column C2 to be printed in
the Session window.

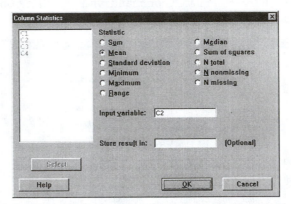

Display I.26: Dialog box for computing column statistics.

If we want to, we can store this result in a constant or column by making an
appropriate entry in the Store result in box. We see from the dialog box that
there are a number of possible statistics that can be computed.

We can also compute statistics row-wise. One difference with column statistics is that these must be stored. For example, suppose we want to compute
the average of the Statistics, Calculus, and Physics marks. The command Çalc
▶ Row Statistics produces the dialog box shown in Display I.27 where we have
placed C2, C3, and C4 into the Input variables box and c6 into the Store result
in box.

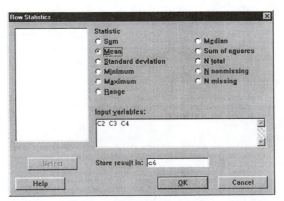

Display I.27: Dialog box for computing row statistics.

It is also possible to compute column statistics using session commands. For example,

```
MTB >mean(c2)
MEAN = 69.900
```

computes the mean of c2. If we want to save the value for subsequent use, then the command

```
MTB >let k1=mean(c2)
```

does this. The general syntax for column statistic commands is

column statistic name(E_1)

where the operation is carried out on the entries in column E_1, and output is written to the screen unless it is assigned to a constant using the **let** command. See Appendix B.2 for a list of all the column statistics available.

Also, for most column statistics there are versions that compute *row statistics,* and these are obtained by placing **r** in front of the column statistic name. For example,

```
MTB >rmean(c2 c3 c4 c6)
```

computes the mean of the corresponding entries in C2, C3, and C4 and places the result in C6. The general syntax for row statistic commands is

row statistic name($E_1 \ldots E_m \ E_{m+1}$)

where the operations are carried out on the rows in columns $E_1, \ldots, E_m$, and the output is placed in column E_{m+1}. See Appendix B.3 for a list of all the row statistics available.

10.4 Comparisons and Logical Operations

Minitab also contains the following comparison and logical operators.

Comparison Operators	Logical Operators
equal to =, **eq**	&, **and**
not equal to <>, **ne**	\, **or**
less than <, **lt**	~, **not**
greater than >, **gt**	
less than or equal to <=, **le**	
greater than or equal to >=, **ge**	

Notice that there are two choices for these operators; for example, use either the symbol >= or the mnemonic **ge.**

The comparison and logical operators are useful when we have simple questions about the worksheet that would be tedious to answer by inspection. This feature is particularly useful when we are dealing with large data sets. For example, suppose that we want to count the number of times the Statistics grade was greater than the corresponding Calculus grade in the marks worksheet. The command Calc ▶ Calculator gives the dialog box shown in Display I.28 where we have put c6 in the Store result in variable box and c2 > c3 in the Expression box. Clicking on the OK button results in the *ith* entry in C6 containing a 1 if the *ith* entry in C2 is greater than the *ith* entry in C3, i.e., the comparison is true, and a 0 otherwise. In this case, C6 contains the entries: 0, 1, 0, 1, 0, 1, 0, 0, 1, 0, which the worksheet in Display I.4 verifies as appropriate. If we use Calc ▶ Calculator to calculate the sum of the entries in C6, we will have computed the number of times the Statistics grade is greater than the Calculus grade.

These operations can also be simply carried out using session commands. For example,

```
MTB >let c6=c2>c3
MTB >let k4=sum(c6)
MTB >print k4
K4 4.00000
```

accomplishes this.

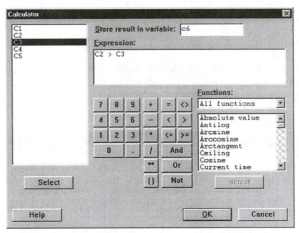

Display I.28: Dialog box for comparisons.

The logical operators combine with the comparison operators to allow more complicated questions to be asked. For example, suppose we wanted to calculate the number of students whose Statistics mark was greater than their Calculus mark and less than or equal to their Physics mark. The commands

```
MTB >let c6=c2>c3 and c2<=c4
MTB >let k4=sum(c6)
MTB >print k4
K4 1.00000
```

accomplish this. In this case, both conditions c2>c3 and c2<=c4 have to be true for a 1 to be recorded in C6. Note that the observation with the missing Physics mark is excluded. Of course, we can also implement this using Calc ▶ Calculator and filling in the dialog box appropriately.

Text variables can be used in comparisons where the ordering is alphabetical. For example,

```
MTB >let c6=c5<''m''
```

puts a 1 in C6 whenever the corresponding entry in C5 is alphabetically smaller than m.

11 Some More Minitab Commands

In this section we discuss some commands that can be very helpful in certain applications. We will make reference to these commands at appropriate places throughout the manual. It is probably best to wait to read these descriptions until such a context arises.

11.1 Coding

The Manip ▶ Code command is used to recode columns. By this we mean that data entries in columns are replaced by new values according to a coding scheme that we must specify. You can recode numeric into numeric, numeric into text, text into numeric, or text into text by choosing an appropriate subcommand. For example, suppose in the marks worksheet we want to recode the grades in C2, C3, and C4 so that any mark in the range 0–39 becomes an F, every mark in the range 40–49 becomes an E, every mark in the range 50–59 becomes a D, every mark in the range 60–69 becomes a C, every mark in the range 70–79 becomes a B, every mark in the range 80–100 becomes an A, and the results are placed in columns C6, C7, and C8, respectively. Then the command Manip ▶ Code ▶ Numeric to Text brings up the dialog box shown in Display I.29. The ranges for the numeric values to be recoded to a common text value are typed in the Original values box, and the new values are typed in the New box. Note that we have used a shorthand for describing a range of data values as discussed in section 7.2. Because the sixth entry of C4 is *, i.e., it is missing, this value is simply recoded as a blank. You can also recode missing values by including * in one of the Original values boxes. If a value in a column is not covered by one of the values in the Original values boxes, then it is simply left the same in the new column.

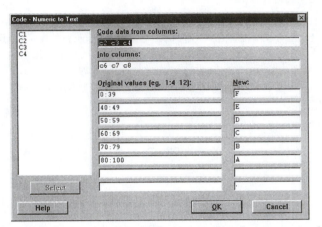

Display I.29: Dialog box for recoding numeric values to text values.

Note that this menu command restricts the number of new code values to 8. The session command **code** allows up to 50 new codes. For example, suppose in the **marks** worksheet we want to recode the grades in C2, C3, and C4 so that any mark in the range 0–9 becomes a 0, every mark in the range 10–19 becomes 10, etc., and the results are placed in columns C6, C7, and C8. The following command

```
MTB >code(0:9) to 0 (10:19) to 10 (20:29) to 20 (30:39) to 30 &
CONT>(40:49) to 40 (50:59) to 50 (60:69) to 60 (70:79) to 70 &
CONT>(80:89) to 80 (90:99) to 90 for C2-C4 put in C6-C8
```

accomplishes this. Note the use of the continuation symbol &, as this is a long command. The general syntax for the **code** command is

$$\textbf{code } (V_1) \text{ to code}_1 \ \dots \ (V_n) \text{ to code}_n \text{ for } E_1 \ \dots \ E_m \text{ put in } E_{m+1} \ \dots \ E_{2m}$$

where V_i denotes a set of possible values and ranges for the values in columns $E_1 \dots E_m$ that are all coded as the number code$_i$, and the results of this coding are placed in the columns $E_{m+1} \dots E_{2m}$, i.e., the recoded E_1 is placed in E_{m+1}, etc.

11.2 Concatenating Columns

The Manip ▶ Concatenate combines two or more text columns into a single text column. For example, if C6 contains m, m, m, f, f, reading first to last entry, and C7 contains to, ta, ti, to, ta, then the entries in the Manip ▶ Concatenate dialog box shown in Display I.30 result in a new text column C8 containing the entries mto, mta, mti, fto, fta.

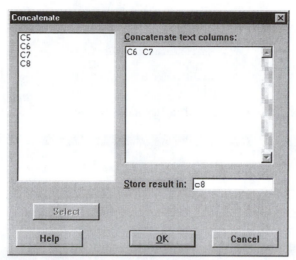

Display I.30: Dialog box for concatenating text columns.

In the session environment, the **concatenate** command is available for this operation. The general syntax of the **concatenate** command is

concatenate $E_1 \ldots E_m$ in E_{m+1}

where $E_1, \ldots, E_m$, are text columns, and E_{m+1} is the target text column.

11.3 Converting Data Types

The Manip ▶ Code ▶ Use Conversion Table command is used to change text data into numeric data and vice versa. As dealing with text data is a bit more difficult in Minitab, we recommend either converting text data to numeric before input or using this command after input to do this.

For example, in the worksheet marks suppose we want to change the gender variable from text, with male and female denoted by m and f, respectively, to a numerical variable with male denoted by 0 and female by 1. To do this, we must first set up a *conversion table*. The conversion table comprises two columns in the worksheet, where one column is text and contains the text values used in the text column, and the second column is numeric and contains the numerical values that you want these changed into. For example, suppose we have entered columns C6 and C7 in the marks worksheet, as shown in Display I.31. The Manip ▶ Code ▶ Use Conversion Table command produces the dialog box shown in Display I.32, where we have indicated that we want to convert the text column C5 into a numeric column and that each m should become a 0 and each f should become a 1.

C6-T	C7
m	0
f	1

Display I.31: Columns c6 and c7 in the marks worksheet as a conversion table.

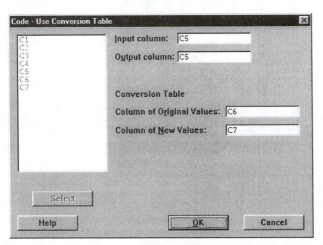

Display I.32: Dialog box for converting text column c5 of the marks worksheet into a numeric column with the conversion table given in columns c6 and c7.

The general syntax for the corresponding session command **convert** is

convert E_1 E_2 E_3 E_4

where E_1, E_2 are the columns containing the conversion table, E_3 is the column to be converted and E_4 is the column containing the converted column.

11.4 History

Minitab keeps a record of the commands you have used and the data you have input in a session. This information can be obtained in the History folder of the Project Manager window. The commands can be copied from wherever they are listed and pasted into the Session window to be reexecuted, so that a number of commands can be executed at once without retyping. These commands can be edited before being executed again. This is very helpful when you have implemented a long sequence of commands and realize that you made an error early on. Note that even if you use the menu commands, a record is kept only of the corresponding session commands.

The **journal** command is available in the Session window if you want to keep a record of the commands in an external file. For example,

```
MTB >journal 'comm1'
Collecting keyboard input(commands and data)in file:
                                                    comm1.MTJ
MTB >read c1 c2 c3
DATA>1 2 3
DATA>end
 1 rows read.
MTB >nojournal
```

puts

```
read c1 c2 c3
1 2 3
end
nojournal
```

into the file `comm1.mtj`. The history is turned off as soon as the **nojournal** command is typed.

11.5 Computing Ranks

Sometimes, we want to compute the *ranks* of the numeric values in a column. The rank r_i of the *ith* value in a column is a value that reflects its relative size in the column. For example, if the *ith* value is the smallest value then $r_i = 1$, if it is the third smallest then $r_i = 3$, etc. If values are the same, i.e., *tied*, then each value receives the average rank. To calculate the ranks of the entries in a column we use the Manip ▶ Rank command. For example, suppose that C6 contains the values 6, 4 , 3, 2, 3, 1. Then the Manip ▶ Rank command brings up the dialog box in Display I.33, which is filled in so that the ranks of the entries in C6 are placed in C7. In this case, the ranks are 6.0, 5.0, 3.5, 2.0, 3.5, and 1.0, respectively.

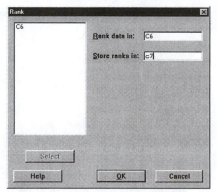

Display I.33: Dialog box for computing ranks.

The syntax of the corresponding session command **rank** is

rank E_1 E_2

where E_1 is the column whose ranks we want to compute, and E_2 is the column that will hold the computed ranks.

11.6 Sorting Data

It often occurs as part of a data analysis that we want to *sort* a column so that its values ascend from smallest to largest or descend from largest to smallest. Note that ordering here could refer to numerical order or alphabetical order, so we also consider ordering text columns. Also, we may want to sort all the rows contained in some subset of the columns in the worksheet *by* a particular column. The Manip ▶ Sort command allows us to carry out these tasks.

For example, suppose that we want to sort the entries in C2 in the **marks** worksheet — the Statistics grades — from smallest to largest and place the sorted values in C6. Then the Manip ▶ Sort command brings up the dialog box shown in Display I.34, where the Sort column(s) box contains the column C2 to be sorted, the Store sorted column(s) In box contains C6, where we will store the sorted column, and C2 is also placed in the Sort by column box. This command results in C6 containing 23, 63, 67, 71, 74, 75, 77, 81, 81, 87. If we had clicked the Descending box, the order of appearance of these values in C6 would have been reversed.

If we had placed another column in the Sort by column box, say C5, then C5 would have been sorted with the values in C2 carried along and placed in C6, i.e., the values in C2 would be sorted *by* the values in C5. So all the Statistics marks of females, in the order they appear in C2 will appear in C6 first and then the Statistics marks of males. For example, replacing C2 by C5 in this box would result in the values in C6 becoming 77, 71, 87, 81, 74, 81, 75, 63, 23, 67. If we fill in the next Sort by column box with another column, say C3, then the values in C2 are sorted first by gender and then within gender by the values in C3.

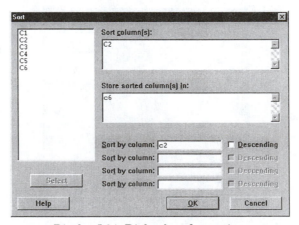

Display I.34: Dialog box for sorting.

The general syntax of the corresponding session command **sort** is

sort E_1 $E_2 \ldots E_m$ $E_{m+1} \ldots E_{2m}$

where E_1 is the column to be sorted, and E_2, ..., E_m are carried along with the results placed in columns E_{m+1}, ..., E_{2m}. Note that this sort can also be accomplished using the **by** subcommand, where the general syntax is

sort E_1 $E_2 \ldots E_m$ $E_{m+1} \ldots E_{2m}$;
by $E_{2m+1} \ldots E_n$.

where now we sort by columns E_{2m+1}, ..., E_n, sorting first by E_{2m+1}, then E_{2m+2}, etc., carrying along E_1, ..., E_m and placing the result in E_{m+1}, ..., E_{2m}. The **descending** subcommand can also be used to indicate which sorting variables we want to use in descending order rather than ascending order.

11.7 Stacking and Unstacking Columns

The M̲anip ▶ St̲ack command is used to literally stack columns one on top of the other or unstack a column into separate columns. For example, in the **marks** worksheet the M̲anip ▶ St̲ack ▶ St̲ack Columns command brings up the dialog box shown in Display I.35, which has been filled in to stack columns C2, C3, and C4 into C6 with the values in C2 first, followed by the values in C3 and then the values in C4. In C7 we have stored an index which indicates that column each value in C6 came from with a 1 every time a value came from C2, a 2 every time a value came from C3, and a 3 every time a value came from C4. It is not necessary to create such an index.

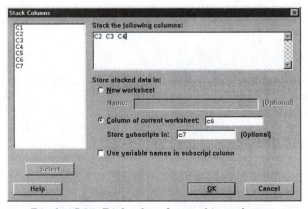

Display I.35: Dialog box for stacking columns.

In the Session window, this same result can be obtained using the **stack** command. The general syntax for the **stack** command is given by

stack $E_1 E_2 \ldots E_m$ into E_{m+1}

where E_1, E_2, ..., E_m denote the columns or constants to be stacked one on top of the other, starting with E_1, and with the result placed in column E_{m+1}. If we want to keep an index of where the values came from, then use the subcommand

subscripts E_{m+2}

which results in index values being stored in column E_{m+2}.

To unstack values in a column by the values in an index column we use the Manip ▶ Unstack command. For example, given the columns C6 and C7 of the **marks** worksheet as described above, the dialog box shown in Display I.36 unstacks C6 into three columns by the values in C7. The three columns are C8, C9, and C10. Note that they are identical to columns C2, C3, and C4, respectively. We must always specify a column containing the subscripts when unstacking a column.

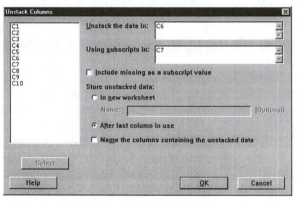

Display I.36: Dialog box for unstacking columns.

The general syntax for the corresponding session command **unstack** is

unstack E_1 into $E_2 \ldots E_m$;
subscripts E_{m+1}.

where E_1 is the column to be unstacked, E_2, ..., E_m are the columns and constants to contain the unstacked column, and E_{m+1} gives the subscripts 1, 2, ... that indicate how E_1 is to be unstacked.

Note that it is also possible to simultaneously unstack blocks of columns. We refer the reader to **help** or Help for information on this.

12 Exercises

1. The following data give the Hi and Low trading prices in Canadian dollars for various stocks on a given day on the Toronto Stock Exchange. Create a worksheet, giving the columns the same variable names, using any of the methods discussed in I.7. Be careful to ensure that the value of the variable `stock` starts with a letter. Print the worksheet to check that you have successfully entered it. Save the worksheet giving it the name `stocks`.

Stock	Hi	Low
ACR	7.95	7.80
MGI	4.75	4.00
BLD	112.25	109.75
CFP	9.65	9.25
MAL	8.25	8.10
CM	45.90	45.30
AZC	1.99	1.93
CMW	20.00	19.00
AMZ	2.70	2.30
GAC	52.00	50.25

2 Retrieve the worksheet `stocks` created in Exercise 1. Change the `Low` value in the stock MGI to 3.95. Calculate the average of the `Hi` and `Low` prices for all the stocks, and save this in a column called `average`. Calculate the average of all the `Hi` prices, and save this in a constant called `avhi`. Similarly, do this for all the `Low` prices, and save this in a constant called `avlo`. Save the worksheet using the same name. Write all the columns out to a file called `stocks.dat`. Print the file `stocks.dat` on your system printer.

3 Retrieve the worksheet created in Exercise 2. Using the Minitab commands discussed in I.10, calculate the number of stocks in the worksheet whose `average` is greater than $5.00 and less than or equal to $45.00.

4 Using the worksheet created in Exercise 2, insert the following stocks at the beginning of the worksheet.

Stock	Hi	Low
CLV	1.85	1.78
SIL	34.00	34.00
AC	14.45	14.05

Delete the variable `average`. Save the worksheet.

5 Using the worksheet created in Exercise 4, sort the stocks into alphabetical order. Calculate the ranks of the individual stocks based on their Hi price, and save the ranking in a new column. Save the worksheet.

6 Using the worksheet created in Exercise 5, calculate the average Hi price of all the stocks beginning in A.

7 Using the worksheet created in Exercise 5, recode all the Low prices in the range $0–9.99 as 1, in the range $10–39.99 as 2, and greater than or equal to $40 as 3, and save the recoded variable in a new column.

8 Using patterned data input, place the values from -10 to 10 in increments of .1 in C1. For each of the values in C1, calculate the value of the quadratic polynomial $2x^2 + 4x - 3$ (i.e., substitute the value in each entry in C1 into this expression) and place these values in C2. Using Minitab commands and the values in C1 and C2, estimate the point in the range from -10 to 10 where this polynomial takes its smallest value and what this smallest value is. Using Minitab commands and the values in C1 and C2 estimate the points in the range from -10 to 10, where this polynomial is closest to 0.

9 Using patterned data input, place values in the range from 0 to 5 using an increment of .01 in C1. Calculate the value of $1 - e^{-x}$ for each value in C1, and place the result in C2. Using Minitab commands, find the largest value in C1 where the corresponding entry in C2 is less than or equal to .5. Note that e^{-x} corresponds to the **exponentiate** command (see Appendix B.1) evaluated at $-x$.

10 Using patterned data input, place values in the range from -4 to 4 using an increment of .01 in C1. Calculate the value of

$$\frac{1}{\sqrt{2\pi}}\, e^{-x^2/2}$$

for each value in C1, and place the result in C2, where $\pi = 3.1415927$. Using **parsums** (see Appendix B.1), calculate the partial sums for C2, and place the result in C3. Multiply C3 times .01. Find the largest value in C1 such that the corresponding entry in C3 is less than or equal to .25.

Part II

Minitab for Data Analysis

Chapter 1

Looking at
Data—Distributions

New Minitab commands discussed in this chapter

Calc ▶ Probability Distributions ▶ Normal
File ▶ Open Graph
File ▶ Save Graph As
Graph ▶ Boxplot
Graph ▶ Chart
Graph ▶ Dotplot
Graph ▶ Histogram
Graph ▶ Pie Chart
Graph ▶ Probability Plot
Graph ▶ Stem-and-Leaf
Graph ▶ Time Series Plot
Manip ▶ Code
Stat ▶ Basic Statistics ▶ Display Descriptive Statistics
Stat ▶ Basic Statistics ▶ Store Descriptive Statistics
Stat ▶ Tables ▶ Tally

This chapter of IPS is concerned with the various ways of presenting and summarizing a data set. By presenting data, we mean convenient and informative methods of conveying the information contained in a data set. There are two basic methods for presenting data, namely graphically and through tabulations. Still, it can be hard to summarize exactly what these presentations are saying about the data. So the chapter also introduces various summary statistics that are commonly used to convey meaningful information in a concise way.

All of these topics can involve much tedious, error prone calculation, if we were to insist on doing them by hand. An important point is that you should almost never rely on hand calculation in carrying out a data analysis. Not only are there many far more important things for you to be thinking about, as the text discusses, but you are also likely to make an error. On the other hand, never blindly trust the computer! Check your results and make sure that they make sense in light of the application. For this, a few simple hand calculations can prove valuable. In working through the problems in IPS, you should try to use Minitab as much as possible, as this will increase your skill with the package and inevitably make your data analyses easier and more effective.

1.1   Tabulating and Summarizing Data

If a variable is categorical, we construct a table using the values of the variable and record the *frequency* (count) of each value in the data and perhaps the *relative frequency* (proportion) of each value in the data as well. These relative frequencies then serve as a convenient summarization of the data.

If the variable is quantitative, we typically *group* the data in some way, i.e., divide the range of the data into nonoverlapping intervals and record the frequency and proportion of values in each interval. Grouping is accomplished using the <u>M</u>anip ▶ C<u>o</u>de command discussed in I.11.1.

If the values of a variable are *ordered*, we can record the *cumulative distribution*, namely the proportion of values less than or equal to each value. Quantitative variables are always ordered but sometimes categorical variables are as well, e.g., when a categorical variable arises from grouping a quantitative variable.

Often, it is convenient with quantitative variables to record the *empirical distribution function*, which for data values $x_1, \ldots, x_n$ and at a value x is given by

$$\hat{F}(x) = \frac{\# \text{ of } x_i \leq x}{n}$$

i.e., $\hat{F}(x)$ is the proportion of data values less than or equal to x. We can summarize such a presentation via the calculation of a few quantities such as the *first quartile*, the *median*, and the *third quartile* or present the *mean* and the *standard deviation*.

We introduce some new commands to carry out the necessary computations using the data shown in Table 1.1. This is data collected by A.A. Michelson and Simon Newcomb in 1882 concerning the speed of light. We will refer to this hereafter as Newcomb's data and place these in the column C1 with the name `time` in the worksheet called `newcomb`.

28	22	36	26	28	28
26	24	32	30	27	24
33	21	36	32	31	25
24	25	28	36	27	32
34	30	25	26	26	25
-44	23	21	30	33	29
27	29	28	22	26	27
16	31	29	36	32	28
40	19	37	23	32	29
-2	24	25	27	24	16
29	20	28	27	39	23

Table 1.1: Newcomb's data..

1.1.1 Tallying Data

The Stat ▶ Tables ▶ Tally command tabulates categorical data. Consider Newcomb's measurements in Table 1.1. These data range from −44 to 40 (use minimum and maximum in Calc ▶ Calculator to calculate these values). Suppose we decide to group these into the intervals $(-50, 0]$, $(0, 20]$, $(20, 25]$, $(25, 30]$, $(30, 35]$, $0(35, 40]$. Next we want to record the frequencies, relative frequencies, cumulative frequencies, and cumulative distribution of this grouped variable. First, we used the Manip ▶ Code ▶ Numeric to Numeric command, as described in I.11.1, to recode the data so that every value in $(-50, 0]$ is given the value 1, every value in $(0, 20]$ is given the value 2, etc., and these values are placed in C2. The dialog box for doing this is shown in Display 1.1.

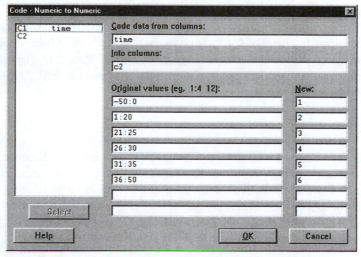

Display 1.1: Dialog box for recoding Newcomb's data.

Next we used the $\underline{\text{S}}$tat ▶ $\underline{\text{T}}$ables ▶ T$\underline{\text{a}}$lly command, with the dialog box shown in Display 1.2,

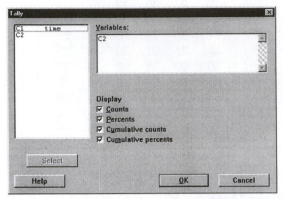

Display 1.2: Dialog box for tallying the variable C2 in the `newcomb` worksheet.

to produce the output

```
C2 Count Percent CumCnt  CumPct
1      2    3.03       2    3.03
2      4    6.06       6    9.09
3     17   25.76      23   34.85
4     26   39.39      49   74.24
5     10   15.15      59   89.39
6      7   10.61      66  100.00
     N= 66
```

in the Session window.

We can also use the $\underline{\text{S}}$tat ▶ $\underline{\text{T}}$ables ▶ T$\underline{\text{a}}$lly command to compute the empirical distribution function of C1 in the `newcomb` worksheet. First, we must sort the values in C1, from smallest to largest, using the $\underline{\text{M}}$anip ▶ $\underline{\text{S}}$ort command described in I.11.6, and then we apply the $\underline{\text{S}}$tat ▶ $\underline{\text{T}}$ables ▶ T$\underline{\text{a}}$lly command to this sorted variable.

The general syntax of the corresponding session command **tally** is

tally $E_1 \ldots E_m$

where E_1, ..., E_m are columns of categorical variables, and the command is applied to each column. If no subcommands are given, then only frequencies are computed, while the subcommands **percents** computes relative frequencies, **cumcnts** computes the cumulative frequency function, and **cumpcts** computes the cumulative distribution of C2. Any of the subcommands can be dropped. For example, the commands

```
MTB >sort c1 c3
MTB >tally c3;
SUBC>cumpcnts;
SUBC>store c4 c5.
```

first use the **sort** command to sort the data in C1 from smallest to largest and place the results in C3. The cumulative distribution is computed for the values in C3 with the unique values in C3 stored in C4 and the cumulative distribution at each of the unique values stored in C5 via the **store** subcommand to **tally.**

1.1.2 Describing Data

The S̲tat ▶ B̲asic Statistics ▶ D̲isplay Descriptive Statistics command is used with quantitative variables to present a numerical summary of the variable values. These values are in a sense a summarization of the empirical distribution of the variable. For example, in the **newcomb** worksheet the dialog box shown in Display 1.3 leads to the output

```
Variable  N  Mean  Median TrMean  StDev SE Mean
time      66 26.21  27.00  27.40  10.75    1.32
Variable Minimum Maximum    Q1    Q3
time      -44.00   40.00 24.00 31.00
```

in the Session window. This provides the count N, the mean, median, trimmed mean **TrMean** (removes lower 5% and upper 5% of the data and averages the rest), standard deviation, standard error of the mean, minimum, maximum, first quartile **Q1**, and third quartile **Q3** of the variable C1. If we want such a summary of a variable by the values of another variable, we check the B̲y variable box and indicate the by variable in the box to the right of this. For example, we might want such a summary for each of the groups we created in II.1.1, and so we would place C2 in this box. Note that a number of summary statistics can also be computed using the Column Statistics discussed in I.10.3.

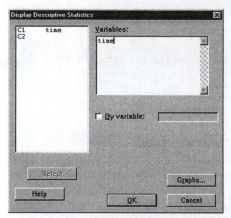

Display 1.3: Dialog box for computing basic descriptive statistics of a quantitative variable.

If we wish to compute some basic statistics and store these values for later use, then the S̲tat ▶ B̲asic Statistics ▶ S̲tore Descriptive Statistics command is available for this. For example, with the **newcomb** worksheet this command leads to the dialog box shown in Display 1.4. Clicking on the S̲tatistics button results in the dialog box of Display 1.5 where we have checked F̲irst quartile, M̲edian, T̲hird quartile, Interquartile range, and N̲ nonmissing as the statistics we want to compute. The result of these choices is that the next available variables in the worksheet contain these values. So in this case, the values of C3–C7 are as depicted in Display 1.6. Note that these variables are now named as well. Note that many more statistics are available using this command.

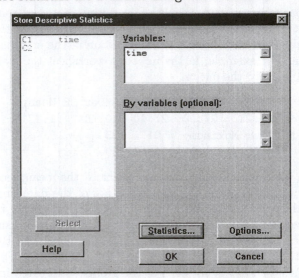

Display 1.4: Dialog box for computing and storing various descriptive statistics.

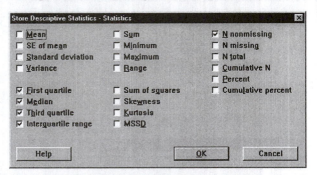

Display 1.5: Dialog box for choosing the descriptive statistics to compute and store.

C3	C4	C5	C6	C7
Q1_1	Median1	Q3_1	IQR1	N1
24	27	31	7	66

Display 1.6: Values obtained for descriptive statistics using dialog boxes in Figures 1.4 and 1.5.

The general syntax of the Session command **describe,** corresponding to S̲tat ▶ B̲asic Statistics ▶ D̲isplay Descriptive Statistics, is

describe $E_1 \ldots E_m$

where E_1, ..., E_m are columns of quantitative variables and the command is applied to each column. A **by** subcommand can also be used. The **stats** command is available in the Session window if we want to store the values of statistics. We refer the reader to **help** for a description of this command.

1.2 Plotting Data in a Graph Window

One of the most informative ways of presenting data is via a plot. There are many different types of plots within Minitab, and which one to use depends on the type of variable you have and what you are trying to learn. In this section we describe how to use the plotting features in Minitab. There are, however, many features of plotting that we will not describe. For example, there are many graphical editing capabilities that allow you to add features, such as titles or legends. Some of these features are accessed via Graph ▶ Layout. We refer the reader to Help for more details on these features.

Each plot in Minitab is made in a *Graph window*. You can make multiple plots and retain each Graph window until you want to delete it simply by clicking the × symbol in the upper right-hand corner. You make any particular Graph window active by clicking in it or by using the Window command. A plot can be saved in an external file in a variety of formats, such as Minitab graph `.mgf`, bitmap `.bmp`, JPEG `.jpg`, etc., using the File ▶ Save Graph As command. If a graph has been saved in the `.mgf` format, it can be reopened using the File ▶ Open Graph command.

1.2.1 Dotplots

The Graph ▶ Dotplot command is used with quantitative variables and produces a plot of each data value as a dot along the x-axis so that you get a general idea of the location of the data and how much scatter there is. Actually, the data is grouped before plotting and multiple observations in a group are stacked over the x-axis. The interval between successive tick (+) marks on the x-axis is divided into 10 equal-length subintervals for the grouping. Typically, one also looks for points that are far from the main scatter of points as these may be identified as *outliers* and, as such, deleted from the data set for subsequent analysis. For example, for the `newcomb` worksheet dialog box in Display 1.7 results in the plot of Display 1.8.

The general syntax of the corresponding Session command **dotplot** is

dotplot $E_1 \ldots E_m$

where E_1, ..., E_m are columns, and a dotplot is produced for each. There are a number of subcommands available. The **same** subcommand ensures the scales of the dotplots are the same for each column. The **by** subcommand allows plotting of a variable by the values of another variable with all plots having the same scale. The **increment** subcommand allows for control of the distance between the tick marks and **start** and **end** allow you to specify where the dotplot should begin and end. For example,

```
MTB >dotplot c1;
SUBC>increment=5;
SUBC>start=20 end=35.
```

puts the tick marks 5 units apart, starts the plot at 20, and ends it at 35, so some points are not plotted in this case.

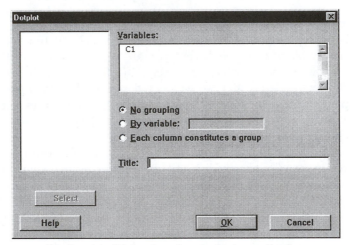

Display 1.7: Dialog box for producing a dotplot.

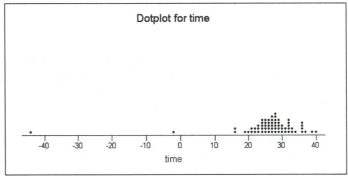

Display 1.8: Dotplot of the Newcomb data.

1.2.2   Stem-and-Leaf Plots

Stem-and-leaf plots are similar to histograms and are produced by the Graph ▶ Stem-and-Leaf command. These plots are also referred to as *stemplots* as in IPS. For example, using this command with the newcomb worksheet produces the output in the Session window

```
Stem-and-leaf of time N = 66
Leaf Unit = 1.0
    1   -4 4
    1   -3
    1   -2
    1   -1
    2   -0 2
    2    0
    5    1 669
  (41)   2 01122333444445555566666777777888888899999
   20    3 0001122222334666679
    1    4 0
```

which is a stem-and-leaf plot of the values in `time`. The first column gives the *depths* for a given stem, i.e., the number of observations on that line and below it or above it, depending on whether or not the observation is below or above the median. The row containing the median is enclosed in parentheses ( ), and the depth is only the observations on that line. If the number of observations is even and the median is the average of values on different rows, then parentheses do not appear. The second column gives the *stems,* as determined by Minitab, and the remaining columns give the ordered *leaves,* where each digit represents one observation. The *Leaf Unit* determines where the decimal place goes after each leaf. So in this example, the first observation is −44.0, while it would be −4.4 if the Leaf Unit were .1. Multiple stem-and-leaf plots can be carried out for a number of columns simultaneously and also for a single variable by the values of another variable.

1.2.3   Histograms

A histogram is a plot where the data are grouped into intervals, and over each such interval a bar is drawn of height equal to the frequency of data values in that interval or of height equal to the relative frequency (proportion) of data values in that interval or of height equal to the *density* of points in that interval, i.e., the proportion of points in the interval divided by the length of the interval. The Graph ▶ Histogram command is used to obtain these plots.

For example, using this command with the `newcomb` worksheet, produces the dialog box shown in Display 1.9. We have placed the variable `time` in the first x box to indicate we want a histogram of this variable. We can produce multiple histograms by placing more variables in the x boxes. To select the type of histogram to plot, we next click on the Options button, which produces the dialog box of Display 1.10. Here, we have selected a density histogram and have specified the intervals to use for grouping the data by specifying the cutpoints −45, −30, −15, 0, 15, 30, 45, which prescribe the intervals $[−45, −30), [−30, −15)$, etc., for the grouping. Alternatively, we could have specified the midpoints of the grouping intervals. The advantage with cutpoints is that subintervals of unequal lengths can be specified. Clicking on the OK buttons in these boxes produces the histogram shown in Display 1.11. As can be seen from the dialog box of Display 1.9, there are a variety of methods for controlling the appearance of the histogram produced, and we refer the reader to the Help button for a description of these.

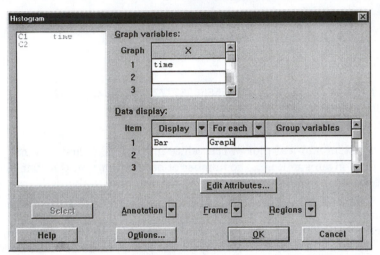

Display 1.9: Dialog box for creating a histogram of the time variable in the `newcomb` worksheet.

Display 1.10: Dialog box for selecting the type of histogram to plot.

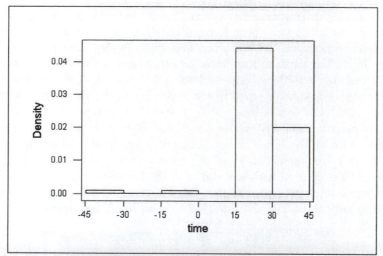

Display 1.11: Density histogram of the time variable in the `newcomb` worksheet.

An important consideration when plotting multiple histograms is to ensure that all the histograms have the same x and y scales so that the plots are visually comparable. This can be accomplished from the dialog box shown in Display 1.9 by Frame ▶ Multiple Graphs and then selecting Same X and same Y.

The session command **histogram** is also available. This has the general syntax

   histogram $E_1 \ldots E_m$

where E_1, ..., E_m correspond to columns. For example, the commands

```
MTB >histogram c1;
SUBC>cutpoints -45 -30 -15 0 15 30 45;
SUBC>density.
```

produce the histogram in Display 1.11 using the **cutpoints** and **density** subcommands. There are also subcommands **midpoints, nintervals,** which specify the number of subintervals, and **frequency** or **percent,** which respectively ensure that the heights of the bar lines equal the frequency and relative frequency of the data values in the interval. Also, the **cumulative** subcommand is available so that the bars represent all the values less than or equal to the endpoint of an interval. The subcommand **same** ensures that multiple histograms all have the same scale.

1.2.4 Boxplots

Boxplots are useful summaries of a quantitative variable and are obtained using the Graph ▶ Boxplot command. Boxplots are used to provide a graphical notion of the location of the data and its scatter in a concise and evocative way. For example, in the `newcomb` worksheet this command produces the dialog box shown in Display 1.12 and the plot in Display 1.13. The line in the center of the box is the median. The line below the median is the first quartile, also called the *lower hinge,* and the line above is third quartile, also called the *upper hinge.* The difference between the third and first quartile, is called the *interquartile range* or IQR. The vertical lines from the hinges are called *whiskers,* and these run from the hinges to *adjacent values.* The adjacent values are given by the greatest value less than or equal to the *upper limit* (the third quartile plus 1.5 times the IQR) and by the least value greater than or equal to the *lower limit* (the first quartile minus 1.5 times the IQR). The upper and lower limits are also referred to as the *inner fences.* The *outer fences* are defined by replacing the multiple 1.5 in the definition of the inner fences by 3.0. Values beyond the outer fences are plotted with a * and are called *outliers.* As with the plotting of histograms, multiple boxplots can be plotted for comparison purposes, and again, it is important to make sure that they all have the same scale.

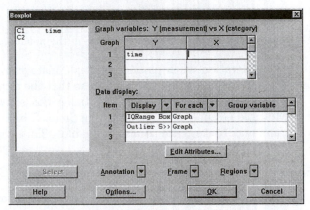

Display 1.12: Dialog box for producing a boxplot of the time variable in the **newcomb** worksheet.

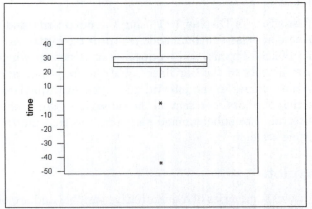

Display 1.13: Boxplot of the time variable in the **newcomb** worksheet.

There is a corresponding session command called **boxplot**. We refer the reader to **help** for more discussion of this command.

1.2.5   Time Series Plots

Often, data are collected sequentially in time. In such a context, it is instructive to plot the values of quantitative variables against time in a time series plot. For this we use the Graph ▶ Time Series Plot command. If we suppose that the data values in `time` of the `newcomb` worksheet were obtained in the order they are listed, then applying this command to that data with the dialog box as in Display 1.14 produces the time plot shown in Display 1.15. Notice that in the Data display box we have specified that the graph should plot a symbol for each point and that the symbols plotted should connect via lines. For example, if we had left out connect, only the points would have been plotted. The lines help to visualize the form of the graph. The symbol plotted is a solid circle but other choices could have been made using the Edit Attributes button. Also, for the Time Scale we have chosen Index, which is just the order in which the observations are listed. If these observations were made at periodic time intervals, there are other possible choices that could be more meaningful.

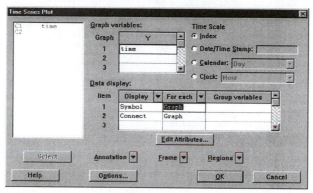

Display 1.14: Dialog box for a time series plot of the variable time from the `newcomb` worksheet.

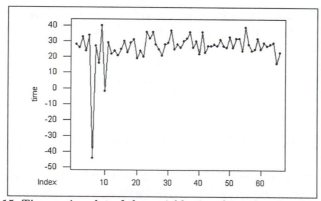

Display 1.15: Time series plot of the variable time from the `newcomb` worksheet.

There is also a corresponding session command **tsplot**. We refer the reader to **help** for more discussion of this.

1.2.6   Bar Charts

It is also possible to produce various charts using the Graph ▶ Chart command. For example, the dialog box shown in Display 1.16 plots a *bar chart* of the variable C2 in the newcomb worksheet. Each distinct value of C1 is plotted along the x-axis simply as a categorical value, not as a quantitative value, and a bar of height equal to the number of times that value occurs in the variable is drawn. A bar chart is a good way to plot categorical variables. There are many possibilities for the types of bar charts drawn, and we refer the reader to the Help button for a discussion of these.

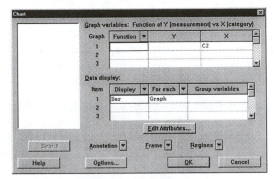

Display 1.16: Dialog box for plotting bar charts.

The corresponding session command is

chart E_1

which produces a bar chart for the values in column E_1.

1.2.7   Pie Charts

A *pie chart* is a disk divided up into wedges where each wedge corresponds to a unique value of a variable, and the area of the wedge is proportional to the relative frequency of the value with which it corresponds. Pie charts can be obtained via Graph ▶ Pie Chart, and there are various features available in the dialog box that can be used to enhance these plots. Pie charts are a common method for plotting categorical variables.

1.3   The Normal Distribution

It is important in statistics to be able to do computations with the normal distribution. The equation of the *density curve* for the normal distribution with mean μ and standard deviation σ is given by

$$\frac{1}{\sqrt{2\pi}}e^{-\frac{1}{2}\left(\frac{z-\mu}{\sigma}\right)^2}$$

where z is a number. We refer to this as the $N(\mu, \sigma)$ density curve. Also of interest is the area under the density curve from $-\infty$ to a number x, i.e., the area between the graph of the $N(\mu, \sigma)$ density curve and the interval $(-\infty, x]$. As noted in IPS, this is a value between 0 and 1. Sometimes, we specify a value p between 0 and 1 and then want to find the point x_p, such that p of the area under the $N(\mu, \sigma)$ density curve lies over $(-\infty, x_p]$. The point x_p is called the *pth percentile* of the $N(\mu, \sigma)$ density curve.

Often, we are given a mean μ and a standard deviation σ and asked to standardize a variable x whose values are in some column, i.e., produce the new variable $z = \frac{x-\mu}{\sigma}$. These arithmetical operations can be carried out using the **let** command as described in I.10.1.

1.3.1   Calculating the Density

Suppose that we want to evaluate the $N(\mu, \sigma)$ density curve at a value x. For this, we use the $\underline{C}$alc ▶ Probability $\_$Distributions ▶ $\underline{N}$ormal command. For example, the dialog box in Display 1.17 indicates that we want to evaluate the $N(10, 1)$ density curve at the value $x = 11.0$.

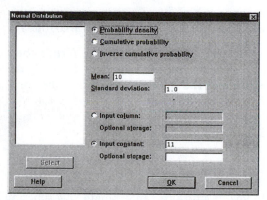

Display 1.17: Dialog box for normal probability calculations.

After clicking on the $\underline{O}$K button the output

```
Normal with mean = 10.0000 and standard deviation = 1.00000
       x              f( x )
   11.0000           0.2420
```

is printed in the Session window, which gives the value as .2420. Sometimes, we will want to evaluate the density curve at every value in a column of values, e.g., when we are plotting this curve. For this we simply click on the radio button Input column and type the relevant column in the associated box.

The general syntax of the corresponding session command **pdf** with the **normal** subcommand is

pdf $E_1 \ldots E_m$ into $E_{m+1} \ldots E_{2m}$;
normal mu $= V_1$ **sigma** $= V_2$.

where E_1, ..., E_m are columns or constants containing numbers and E_{m+1}, ..., E_{2m} are the columns or constants that store the values of the $N(\mu, \sigma)$ density curve at these numbers and $V_1 = \mu$ and $V_2 = \sigma$. If no storage is specified, then the values are printed. For example, if we want to compute the $N(-.5, 1.2)$ density curve at every value between -3 and 3 in increments of .01, the commands

```
MTB >set c1
DATA>-3:3/.01
DATA>end
MTB >pdf c1 c2;
SUBC>normal mu=-.5 sigma=1.2.
```

put the values between -3 and 3 in increments of .01 in C1 using the **set** command. The **pdf** command with the **normal** subcommand calculates the $N(-.5, 1.2)$ density curve at each of these values and puts the outcomes in the corresponding entries of C2. If we plot C2 against C1, we will have a plot of the density curve of this distribution. For this, we use the scatterplot facilities in Minitab as discussed in II.3. Note that with the **normal** subcommand we must also specify the mean and the standard deviation via **mu** and **sigma**.

1.3.2   Calculating the Distribution Function

Suppose that we want to evaluate the area under $N(\mu, \sigma)$ density curve over the interval $(-\infty, x]$. This is the value of the cumulative distribution function of the $N(\mu, \sigma)$ distribution at the value x. For this, we use the Çalc ▶ Probability Ðistributions ▶ Normal as well, but in this case in the dialog box of Display 1.17 we select Çumulative probability instead. Making this change in the dialog box of Display 1.17, we get the output

```
      x          P( X <= x )
  11.0000          0.8413
```

in the Session window. Again, we can evaluate this function at a single point or at every value in a variable.

The general syntax of the corresponding Session command **cdf** command with the **normal** subcommand is

cdf $E_1 \ldots E_m$ into $E_{m+1} \ldots E_{2m}$;
normal mu $= V_1$ **sigma** $= V_2$.

where E_1, ..., E_m are columns or constants containing numbers and E_{m+1}, ..., E_{2m} are the columns or constants that store the values of the area under $N(\mu, \sigma)$ density curve over the interval from $-\infty$ to these numbers and $V_1 = \mu$ and $V_2 = \sigma$. If no storage is specified, the values are printed.

1.3.3   Calculating the Inverse Distribution Function

Suppose that we want to evaluate percentiles for the $N(\mu, \sigma)$ density curve. Again, we use the Çalc ▶ Probability Ðistributions ▶ Normal command, but in this case, in the dialog box of Display 1.17 we select <u>I</u>nverse cumulative probability instead. Making this change in the dialog box of Display 1.17 and replacing 11 by .75 — recall that the argument to this function must be between 0 and 1 — we get the output

```
P( X <= x )          x
   0.7500        10.6745
```

in the Session window. This indicates that the area to the left of 10.6745 underneath the $N(-.5, 1.2)$ density curve is .75.

The general syntax of the corresponding session command **invcdf** with the **normal** subcommand is

invcdf $E_1 \ldots E_m$ into $E_{m+1} \ldots E_{2m}$;
normal mu $= V_1$ **sigma** $= V_2$.

where E_1, ..., E_m are columns or constants containing numbers between 0 and 1 and E_{m+1}, ..., E_{2m} are the columns or constants that store the values of the percentiles of the $N(\mu, \sigma)$ density curve at these numbers and where $V_1 = \mu$ and $V_2 = \sigma$. If no storage is specified, then the values are printed.

1.3.4   Normal Probability Plots

Some statistical procedures require that we assume that values for some variables are a sample from a normal distribution. A *normal probability plot* is a *diagnostic* that checks for the reasonableness of this assumption. To create such a plot, we use the <u>G</u>raph ▶ Probability Plot command. For example, using this command on the **newcomb** worksheet we get the dialog box in Display 1.18 where we have placed **time** in the <u>V</u>ariables box. Clicking on the <u>O</u>K button produces the plot in Display 1.19. The normal probability plot is given by the dark dotted curve. The plot also contains other information and further output is printed in the Session window. Of course, the plot should be like a straight line and it is not in this case.

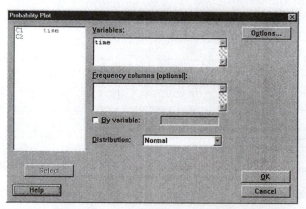

Display 1.18: Dialog box for producing normal probability plots.

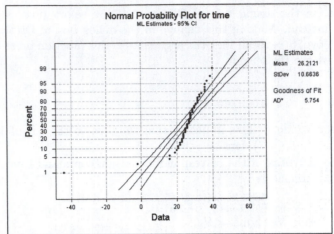

Display 1.19: Normal probability plot of the time variable in the newcomb worksheet.

The session commands
```
MTB >nscores c1 c3
MTB >plot c3*c1
```

produce a normal probability plot like that shown in Display 2.3. The **plot** command will be discussed much more extensively in II.3. The **nscores** (*normal scores*) command relies on some concepts that are beyond the level of this course so we do not discuss this further.

1.4   Exercises

When the data for an exercise come from an exercise in IPS, the IPS exercise number is given in parentheses ( ). All computations in these exercises are to be carried out using Minitab, and the exercises are designed to ensure that you have a reasonable understanding of the Minitab material in this chapter. Generally, you should be using Minitab to do all the computations and plotting required for the problems in IPS.

1. Using Newcomb's measurements in Table 1.1, create a new variable by grouping these values into three subintervals [−50, 0), [0, 20), [20, 50). Calculate the frequency distribution, the relative frequency distribution, and the cumulative distribution of this ordered categorical variable.

2. (1.21) Use Minitab to print the empirical distribution function. From this, determine the first quartile, median, and third quartile. Also, use the empirical distribution function to compute the 10th and 90th percentiles.

3. Use Minitab to produce the stemplot of Example 1.4 of IPS.

4. Use Minitab to produce the time plot of Example 1.5 of IPS.

5. (1.29) Use Minitab commands for the stemplot and the time plot. Use Minitab commands to compute a numerical summary of this data, and justify your choices.

6. (1.30) Transform the data in this problem by subtracting 5 from each value and multiplying by 10. Calculate the means and standard deviations, using any Minitab commands, of both the original and transformed data. Compute the ratio of the standard deviation of the transformed data to the standard deviation of the original data. Comment on this value.

7. (1.30) Transform this data by multiplying each value by 3. Compute the ratio of the standard deviation to the mean (called the *coefficient of variation*) for the original data and for the transformed data. Justify the outcome.

8. For the $N(6, 1.1)$ density curve, compute the area between the interval $(3, 5)$ and the density curve. What number has 53% of the area to the left of it for this density curve?

9. Use Minitab commands to verify the 68-95-99.7 rule for the $N(2, 3)$ density curve.

10. Calculate and store the values of the $N(0, 1)$ density curve at each value in $[-3, 3]$ using an increment of .01. Put the values in the interval $[-3, 3]$ in C1 and the values of the density curve in C2. Using the command `plot C2*C1`, plot the density curve. Comment on the shape of this curve.

11. Use Minitab commands to make the normal quantile plots presented in Figures 1.31 and 1.32 of IPS.

Chapter 2

Looking at Data—Relationships

New Minitab commands discussed in this chapter

Graph ▶ Plot
Stat ▶ Basic Statistics ▶ Correlation
Stat ▶ Regression ▶ Fitted Line Plot
Stat ▶ Regression ▶ Regression

In this chapter, Minitab commands are described that permit the analysis of relationships among two variables. The methods are different depending on whether or not both variables are quantitative, both variables are categorical, or one is quantitative and the other is categorical. This chapter considers relationships between two quantitative variables with the remaining cases discussed in later chapters. Graphical methods are very useful in looking for relationships among variables, and we examine various plots for this.

2.1   Scatterplots

A *scatterplot* of two quantitative variables is a useful technique when looking for a relationship between two variables. By a scatterplot we mean a plot of one variable on the y-axis against the other variable on the x-axis. For example, consider Example 2.4 in IPS, where we are concerned with the relationship between the length of the femur and the length of the humerus in an extinct species. Suppose that we have input the data so that length of the femur measurements are in C1, which has been named **femur**, and the length of the humerus measurements are in C2, which has been named **humerus**, of the worksheet **archaeopteryx**. The command Graph ▶ Plot produces the dialog box of Display 2.1, where we have placed **femur** into the first box for the y variable and `humerus` in the first box for the x variable. This produces the plot shown in Display 2.2. Note that we could alter the plotting symbol using the dialog box that appears when we click on the Edit Attributes box. Using the dialog box that appears when you click on the Annotation button, it is possible to give the plot a title, label plotted points, etc. Using the dialog box that appears when you click on the Frame button, you can change the labels on the axes. Rather than just plotting the points in a scatterplot, you can add *connection lines* (join the points with lines), add *projection lines* (drop a line from each point to the x-axis), and add *areas* (fill in the area under a polygon joining the points). Also, you can employ the scatterplot smoother *lowess* to plot a piecewise linear continuous curve through the scatter of points. These features are available via Graph ▶ Plot ▶ Display. There are a number of other features that allow you to control the appearance of the plot.

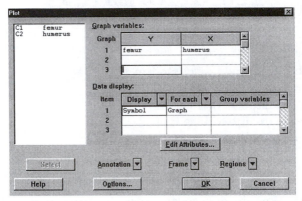

Display 2.1: Dialog box for producing a scatterplot.

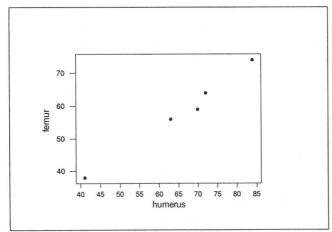

Display 2.2: Scatter plot of femur length (C1) versus humerus length (C2) of Example 2.4 in IPS.

It is also possible to have multiple scatterplots on the same plot. For example, suppose that C3 in the `archaeopteryx` worksheet contains the natural log of the `femur` variable. We obtained the plot of Display 2.3 by adding another pair of variables to the second Graph variables box as in Display 2.1 with C3 as the y variable and `humerus` as the x variable. To put these scatterplots on the same plot use Frame ▶ Multiple Graphs and click on the Overlay graphs on the same page radio button.

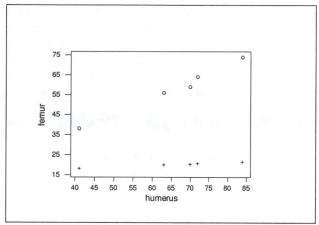

Display 2.3: Multiple scatterplots in the same plot.

The technique of *brushing* is available after obtaining the plot to see which observations (rows) the points correspond to. This is helpful in identifying the points that correspond to outliers. Brushing is accessed from the toolbar just below the menu bar by clicking on the brush when the Graph window is active.

The corresponding session command is **plot.** For example,

```
MTB > plot femur*humerus
```

produces the plot of Display 2.2. Note that the first variable is plotted along the y-axis, and the second variable is plotted along the x-axis. There are various subcommands that can be used with **plot,** and we refer the reader to Help for a description of these.

There are a number of additional plots available in Minitab that are related to the scatterplot. For example, a *marginal plot* of two variables is a scatterplot of one variable against the other where in addition histograms, dotplots or boxplots are plotted along the sides of the scatterplot for each variable. These are available via the menu command Graph ▶ Marginal Plot. *Draftsman plots* allow you to produce a number of scatterplots in a rectangular array so that they can be compared. For example, you may want to plot C1 against C3, C2 against C3, C1 against C4, and C2 against C4 and see all of these in a common plot. This capability is available via the menu command Graph ▶ Draftsman Plot and filling in the dialog box. *Matrix plots* provide a mechanism for placing a number of scatterplots in a rectangular array or matrix so that they can be directly compared or examined for relationships. Matrix plots are available via the command $\underline{G}$raph ▶ $\underline{M}$atrix Plot. Also three-dimensional scatterplots are available via $\underline{G}$raph ▶ $\underline{3}$D Plot and contour plots via $\underline{G}$raph ▶ $Co\underline{n}$tour Plot.

2.2   Correlations

While a scatterplot is a convenient graphical method for assessing whether or not there is any relationship between two variables, we would also like to assess this numerically. The *correlation coefficient* provides a numerical summarization of the degree to which a linear relationship exists between two quantitative variables, and this can be calculated using the $\underline{S}$tat ▶ $\underline{B}$asic Statistics ▶ $\underline{C}$orrelation command. For example, applying this command to the `femur` and `humerus` variables of the worksheet `archaeopteryx`, i.e., the data of Example 2.4 in IPS and depicted in Display 2.2, we obtain the output

```
Pearson correlation of femur and humerus = 0.994
P-Value = 0.001
```

in the Session window. For now, we ignore the number recorded as `P-Value`.

The general syntax of the corresponding session command **correlate** is given by

correlate $E_1 \ldots E_m$

where E_1, ..., E_m are columns corresponding to numerical variables, and a correlation coefficient is computed between each pair. This gives $m(m-1)/2$ correlation coefficients. The subcommand **nopvalues** is available if you want to suppress the printing of P-values.

2.3   Regression

Regression is another technique for assessing the strength of a linear relationship existing between two variables and it is closely related to correlation. For this, we use the $\underline{S}$tat ▶ $\underline{R}$egression command.

As noted in IPS, the regression analysis of two quantitative variables involves computing the least-squares line $y = a + bx$, where one variable is taken to be the response variable y and the other is taken to be the explanatory variable x. Note that the least squares line is different depending upon which choice is made. For example, for the data of Example 2.4 in IPS and plotted in Display 2.2 letting `femur` be the response and `humerus` be the predictor or explanatory variable, the $\underline{S}$tat ▶ $\underline{R}$egression ▶ $\underline{R}$egression command leads to the dialog box of Display 2.4, where we have made the appropriate entries in the R$\underline{e}$sponse and Predi$\underline{c}$tors boxes. Clicking on the $\underline{O}$K button leads to the output of Display 2.5 being printed in the Session window. This gives the least-squares line as $y = 3.70 + .826x$, i.e., $a = 3.70$ and $b = .826$, which we also see under the `Coef` column in the first table. In addition, we obtain the value of the square of the correlation coefficient, also known as the *coefficient of determination*, as `R-Sq = 98.8%`. We will discuss the remaining output from this command in II.10.

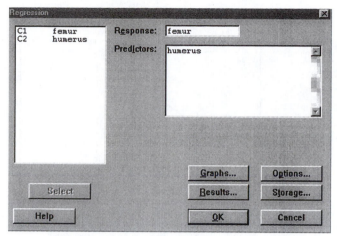

Display 2.4: Dialog box for a regression analysis.

Regression Analysis: femur versus humerus

```
The regression equation is
femur = 3.70 + 0.826 humerus

Predictor        Coef      SE Coef           T        P
Constant        3.701        3.497        1.06    0.368
humerus       0.82574      0.05180       15.94    0.001

S = 1.646      R-Sq = 98.8%     R-Sq(adj) = 98.4%

Analysis of Variance

Source             DF           SS          MS         F        P
Regression          1       688.67      688.67    254.10    0.001
Residual Error      3         8.13        2.71
Total               4       696.80
```

Display 2.5: Output from the dialog box of Display 2.4.

It is very convenient to have a scatterplot of the points together with the least-squares line. This can be accomplished using the Stat ▶ Regression ▶ Fitted Line Plot command. Filling in the dialog box for this command as in Display 2.4 produces the output in the Session window of Display 2.5 together with the plot of Display 2.6.

There are some additional quantities that are often of interest in a regression analysis. For example, you may wish to have the fitted values $\hat{y} = a + bx$ at each x value printed as well as the residuals $y - \hat{y}$. Clicking on the Results button in the dialog box of Display 2.4 and filling in the ensuing dialog box as in Display 2.7 results in these quantities being printed in the Session window as well as the output of Display 2.5.

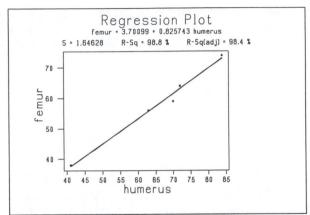

Display 2.6: Scatterplot of femur versus humerus in the archaeopteryx worksheet
together with the least-squares line.

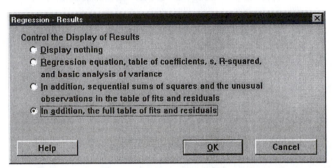

Display 2.7: Dialog box for controlling output for a regression analysis.

You will probably want to keep these values for later work. In this case, clicking
on the Storage button of Display 2.4 and filling in the ensuing dialog box as
in Display 2.8 results in these quantities being saved in the next two available
columns — in this case, C3 and C4 — with the names `resl1` and `fits1` for the
residuals and fits, respectively.

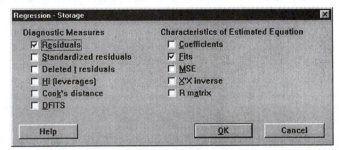

Display 2.8: Dialog box for storing various quantities computed as part of a
regression analysis.

Even more likely is that you will want to plot the residuals as part of assessing
whether or not the assumptions that underlie a regression analysis make sense

in the particular application. For this, click on the Graphs button in the dialog box of Display 2.4. The dialog box of Display 2.9 becomes available. Notice that we have requested that the *standardized residuals* — each residual divided by its standard error — be plotted, and this plot appears in Display 2.10. All the standardized residuals should be in the interval $(-3, 3)$, and no pattern should be discernible. In this case, this residual plot looks fine. From the dialog box of Display 2.9, we see that there are many other possibilities for residual plots.

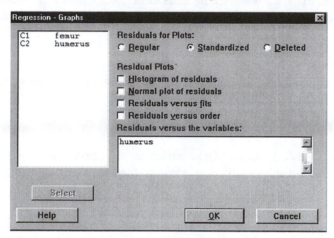

Display 2.9: Dialog box for selecting various residual plots as part of a regression analysis.

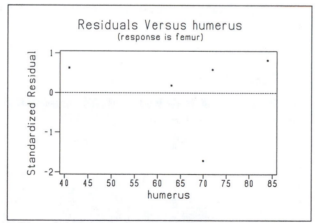

Display 2.10: Plot of the standardized residuals versus humerus after regressing femur against humerus in the archaeopteryx worksheet.

The corresponding session command is given by **regress,** and by using the subcommands **pfits, residual,** and **sresidual** we can calculate and store *fitted values, residuals,* and *standardized residuals,* respectively. For example,

```
MTB > regress c1 1 c2;
SUBC> fits c3;
SUBC> residuals c4;
SUBC> sresiduals c5.
```

gives the output of Display 2.5 and also stores the fitted values in C3, stores the residuals $y - \hat{y}$ in C4, and stores the standardized residuals in C5. Note that the 1 in **regress c1 1 c2** refers to the number of predictors we are using to predict the response variable. To plot the standardized residuals against **humerus**, we use

```
MTB > plot c5*c2
```

which results in a plot like Display 2.10 but with different labels on the x axis.

2.4   Transformations

Sometimes, transformations of the variables are appropriate before we carry out a regression analysis. This is accomplished in Minitab using the Calc ▶ Calculator command and the arithmetical and mathematical operations discussed in I.10.1 and I.10.2. In particular, when a residual plot looks bad, sometimes this can be fixed by transforming one or more of the variables using a simple transformation, such as replacing the response variable by its logarithm or something else. For example, if we want to calculate the cube root — i.e., $x^{1/3}$ — of every value in C1 and place these in C2, we use the Calc ▶ Calculator command and the dialog box as depicted in Display 2.11. Alternatively, we could use the session command **let** as in

```
MTB > let c2=c1**(1/3)
```

which produces the same result.

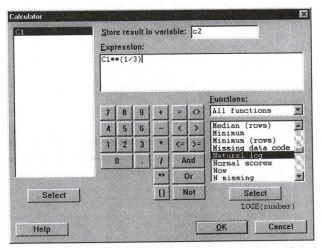

Display 2.11: Dialog box for calculating transformations of variables.

2.5 Exercises

When the data for an exercise come from an exercise in IPS, the IPS exercise number is given in parentheses ( ). All computations in these exercises are to be carried out using Minitab, and the exercises are designed to ensure that you have a reasonable understanding of the Minitab material in this chapter. Generally, you should be using Minitab to do all the computations and plotting required for the problems in IPS.

1. (2.10) Calculate the least-squares line and make a scatterplot of Fuel used against Speed together with the least-squares line. Plot the standardized residuals against Speed. What is the squared correlation coefficient between these variables?

2. (2.11) Make a scatterplot of Rate against Mass where the points for different Sexes are labeled differently (use Minitab for the labeling, too) and with the least-squares line on it. Hint: Make use of the stack command discussed in I.11.7.

3. Place the values 1 through 100 with an increment of .1 in C1 and the square of these values in C2. Calculate the correlation coefficient between C1 and C2. Multiply each value in C1 by 10, add 5, and place the results in C3. Calculate the correlation coefficient between C2 and C3. Why are these correlation coefficients the same?

4. Place the values 1 through 100 with an increment of .1 in C1 and the square of these values in C2. Calculate the least-squares line with C2 as response and C1 as explanatory variable. Plot the standardized residuals. If you see such a pattern of residuals what transformation, might you use to remedy the problem?

5. (2.54) For the data in this problem, numerically verify the algebraic relationship that exists between the correlation coefficient and the slope of the least-squares line.

6. For Example 2.17 in IPS, calculate the least-squares line and reproduce Display 2.21. Calculate the sum of the residuals and the sum of the squared residuals and divide this by the number of data points minus 2. Is there anything you can say about what these quantities are equal to in general?

7. (2.62) Use Minitab to do all the calculations in this problem.

8. Place the values 1 through 10 with an increment of .1 in C1, and place $\exp(-1 + 2x)$ of these values in C2. Calculate the least-squares line using C2 as the response variable, and plot the standardized residuals against C1. What transformation would you use to remedy this residual plot? What is the least-squares line when you carry out this transformation?

Chapter 3

Producing Data

New Minitab commands discussed in this chapter

Çalc ▶ Set Base
Çalc ▶ Random Data

This chapter is concerned with the collection of data, perhaps the most important step in a statistical problem, as this determines the quality of whatever conclusions are subsequently drawn. A poor analysis can be fixed if the data collected are good by simply redoing the analysis. But if the data have not been appropriately collected, then no amount of analysis can rescue the study. We discuss Minitab commands that enable you to generate samples from populations and also to randomly allocate treatments to experimental units.

Minitab uses computer algorithms to mimic randomness. Still, the results are not truly random. In fact, any simulation in Minitab can be repeated, with exactly the same results being obtained, using the Çalc ▶ Set Base command. For example, in the dialog box of Display 3.1 we have specified the base, or seed, random number as 1111089. The base can be any integer. When you want to repeat the simulation, you give this command, with the same integer. Provided you use the same simulation commands, you will get the same results. This can also be accomplished using the session command **base** V, where V is an integer.

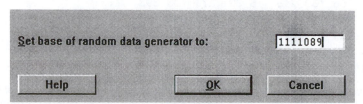

Display 3.1: Dialog box for setting base or seed random number.

3.1 Generating a Random Sample

Suppose that we have a large population of size N and we want to select a sample of $n < N$ from the population. Further, we suppose that the elements of the population are ordered, i.e., we have been able to assign a unique number $1, \ldots, N$ to each element of the population. To avoid selection biases, we want this to be a *random sample,* i.e., every subset of size n from the population has the same "chance" of being selected. As discussed in IPS, this implies that we generate our sample so that every subset of size n in the population has the same chance of being chosen. We can do this physically by using some simple random system, such as chips in a bowl or coin tossing. We could also use a table of random numbers, or, more conveniently, we can use computer algorithms that mimic the behavior of random systems.

For example, suppose there are 1000 elements in a population, and we want to generate a sample of 50 from this population without replacement. We can use the Calc ▶ Random Data ▶ Sample from Columns command to do this. For example, suppose we have labeled each element of the population with a unique number in $1, 2, \ldots, 1000$, and, further, we have put these numbers in C1 of a worksheet. The dialog box of Display 3.2 results in a random sample of 50 being generated without replacement from C1 and stored in C2.

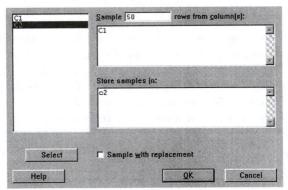

Display 3.2: Dialog box for generating a random sample without replacement.

Printing this sample gives the output

```
MTB > print c2
C2
  441 956   87 736 185 515 883 957 690
  438 205 760 246   16 321 371 493 393
  538 348   70   54 362 492 182 841 287
  277 112 610 890 503 332 413 886 798
  764 584 566 495 547 488 206 557 263
  414 613 618 685 864
```

in the Session window. So now we go to the population and select the elements labeled 441, 956, 87, etc. The algorithm that underlies this command is such that we can be confident that this sample of 50 is like a random sample.

The general syntax of the corresponding session command **sample** is

sample V $E_1 \ldots E_m$ put into $E_{m+1} \ldots E_{2m}$

where V is the sample size n and V rows are sampled from the columns E_1, ..., E_m and stored in columns E_{m+1}, ..., E_{2m}. If we wanted to sample with replacement — i.e., after a unit is sampled, it is placed back in the population so that it can possibly be sampled again — we use the **replace** subcommand. Of course, for simple random sampling, we do not use the **replace** subcommand. Note that the columns can be numeric or text.

Sometimes we want to generate *random permutations*, i.e., $n = N$, and we are simply reordering the elements of the population. For example, in experimental design, suppose we have $N = n_1 + \cdots + n_k$ experimental units and k treatments, and we want to allocate n_i applications of treatment i. Suppose further that we want all possible such applications to be equally likely. Then we generate a random permutation $(l_1, \ldots, l_N)$ of $(1, \ldots, N)$ and allocate treatment 1 to those experimental units labeled $l_1, \ldots, l_{n_1}$, allocate treatment 2 to those experimental units labeled $l_{n_1+1}, \ldots, l_{n_1+n_2}$, etc. For example, if we have 30 experimental units and 3 treatments and we want to allocate 10 experimental units to each treatment, placing the numbers $1, 2, \ldots, 30$ in C1 and using the Calc ▶ Random Data ▶ Sample from Columns command as in the dialog box of Display 3.2, but with 30 in the Sample box, generates a random permutation of $1, 2, \ldots, 30$ in C2. Implementing this gives us the random permutation

```
MTB > print c2
C2
  13   7 26   8 22 23 28 17   3 25
   9   2 14 29 15 18   6 11 16   5
  12 27   4 30 20 24   1 19 21 10
```

and for the treatment allocation you can read the numbers row-wise or column-wise, as long as you are consistent. Row-wise is probably best, as this is how the numbers are stored in C2, and so you can always refer back to C2 (presuming you save your worksheet) if you get mixed up.

The above examples show how to directly generate a sample from a population of modest size. But what happens if the population is huge or it is not convenient to label each unit with a number? For example, suppose we have a population of size 100,000 for which we have an ordered list and we want a sample of size 100. In this case more sophisticated techniques need to be used, but simple random sampling can still typically be accomplished (see Exercise 3.3 for a simple method that works in some contexts).

Simple random sampling corresponds to sampling without replacement, i.e., after we randomly select an element from the population, we do not return it to the population before selecting the next sample element. Sampling with replacement corresponds to replacing each sample element in the population after selecting it and recording only the element that was obtained. So at each selection, every element has the same chance of being selected, and an element may appear more than once in the sample. Notice that we can also sample with replacement if we check the Sample with replacement box in the dialog box of Display 3.2.

3.2   Sampling from Distributions

Once we have generated a sample from a population, we measure various attributes of the sampled elements. For example, if we were sampling from a population of humans, we might measure each sampled unit's height. The height for the sample unit is now a random variable that follows the height distribution in the population from which we are sampling. For example, if 80% of the people in the population are between 4.5 feet and 6 feet, then under *repeated sampling* of an element from the population (with replacement) in the long run, 80% of the sampled units will have their heights in this range.

Sometimes, we want to sample directly from this population distribution, i.e., generate a number in such a way that under repeated sampling in the long run the proportion of values falling in any range agrees with that prescribed by the population distribution. Of course, we typically don't know the population distribution, as this is what we want to find out about in a statistical investigation. Still, there are many instances where we want to pretend that we do know it and simulate from this distribution, e.g., perhaps we want to consider the effect of various choices of population distribution on the sampling distribution of some statistic of interest.

There are computer algorithms that allow us to do this for a variety of distributions. In Minitab, this is accomplished using the Calc ▶ Random Data command. For example, suppose that we want to simulate the tossing of a fair coin (a coin where head and tail are equally likely as outcomes). The Calc ▶ Random Data ▶ Bernoulli command together with the dialog box of Display 3.3 generates a sample of 100 from the *Bernoulli*(.5) distribution and places these values in C1. A random variable has a *Bernoulli*(p) distribution if the probability the variable equals 1 — success — is p and the probability the variable equals 0 — failure — is $1 - p$. So to generate a sample of n from the *Bernoulli*(p) distribution, we put n in the Generate box and p in the Probability of success box. In such a case, we are simulating the tossing of a coin that produces a head on a single toss with probability p, i.e., the long-run proportion of heads that we observe in repeated tossing is p. Note that we can generate m samples of size n by putting m distinct columns in the Store in column(s) box.

Often, a normal distribution with some particular mean and standard deviation is considered a reasonable assumption for the distribution of a measurement in a population. For example, the Calc ▶ Random Data ▶ Normal command together with the dialog box of Display 3.4 generates a sample of 200 from the $N(5.2, 1.3)$ distribution and places this sample in C1. To generate a sample of n from the $N(\mu, \sigma)$ distribution, we put n in the Generate box, μ in the Mean box, and σ in the Standard deviation box.

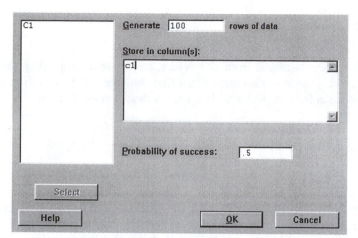

Display 3.3: Dialog box for generating a sample of 100 from the *Bernoulli*(.5) distribution.

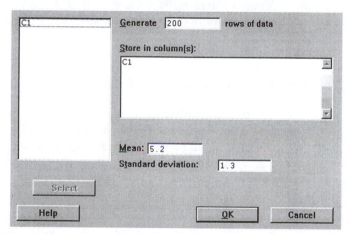

Display 3.4: Dialog box for generating a sample of 200 from a $N(5.2, 1.3)$ distribution.

The general syntax of the corresponding session command **random** is

random V into $E_1 \ldots E_m$

and this puts a sample of size V into each of the columns E_1, ..., E_m, according to the distribution specified by the subcommand. For example,

```
MTB > random 100 c1;
SUBC> bernoulli .5.
```

simulates the tossing of a fair coin 100 times and places the results in C1 using the **bernoulli** subcommand. If no subcommand is provided, this distribution is taken to be the $N(0, 1)$ distribution. The command

```
MTB > random 200 c1;
SUBC> normal mu=2.1 sigma=3.3.
```

generates a sample of 200 from the $N(2.1, 3.3)$ distribution using the **normal** subcommand. There are a number of other subcommands specifying distributions, and we refer the reader to **help** for a description of these.

3.3 Exercises

When the data for an exercise come from an exercise in IPS, the IPS exercise number is given in parentheses ( ). All computations in these exercises are to be carried out using Minitab, and the exercises are designed to ensure that you have a reasonable understanding of the Minitab material in this chapter. Generally, you should be using Minitab to do all the computations and plotting required for the problems in IPS.

If your version of Minitab places restrictions such that the value of the simulation sample size N requested in these problems is not feasible, then substitute a more appropriate value. Be aware, however, that the accuracy of your results is dependent on how large N is.

1. (3.13) Generate a random permutation of the names using Minitab.

2. (3.32) Use the Manip ▶ Sort command described in I.11.6 to order the subjects by weight. Use the values 1–5 to indicate five blocks of equal length in a separate column, and then use the Manip ▶ Unstack command described in I.11.7 to put the blocks in separate columns. Generate a random permutation of each block.

3. Use the following methodology to generate a sample of 20 from a population of 100,000. First, put the values 0–9 in each of C1–C5. Next, use sampling with replacement to generate 50 values from C1, and put the results in C6. Do the same for each of C2–C5 and put the results in C7–C10 (don't generate from these columns simultaneously). Create a single column of numbers using the digits in C6–C10 as the digits in the numbers. Pick out the first unique 20 entries as labels for the sample. If you do not obtain 20 unique values, repeat the process until you do. Why does this work?

4. Suppose you wanted to carry out stratified sampling where there are 3 strata, with the first stratum containing 500 elements, the second stratum containing 400 elements, and the third stratum containing 100 elements. Generate a stratified sample with 50 elements from the first stratum, 40 elements from the second stratum, and 10 elements from the third stratum. When the strata sample sizes are the same proportion of the total sample size as the strata population sizes are of the total population size this is called _proportional sampling_.

5. Suppose we have an urn containing 100 balls with 20 labeled 1, 50 labeled 2, and 30 labeled 3. Using sampling with replacement, generate a sample of size 1000 from this distribution employing the Calc ▶ Random Data command to generate the sample directly from the relevant population distribution. Use the Stat ▶ Tables ▶ Cross Tabulation command to record the proportion of each label in the sample.

6. Carry out a simulation study with $N = 1000$ of the sampling distribution of $\hat{p}$ for $n = 5, 10, 20$ and for $p = .5, .75, .95$. In particular, calculate the empirical distribution functions and plot the histograms. Comment on your findings.

7. Carry out a simulation study with $N = 2000$ of the sampling distribution of the sample standard deviation when sampling from the $N(0, 1)$ distribution based on a sample of size $n = 5$. In particular, plot the histogram using cutpoints 0, 1.5, 2.0 2.5, 3.0 5.0. Repeat this for the sample coefficient of variation (sample standard deviation divided by the sample mean) using the cutpoints -10, -9, ..., 0, ..., 9, 10. Comment on the shapes of the histograms relative to an $N(0, 1)$ density curve.

Chapter 4

Probability: The Study of Randomness

In this chapter the concept of probability is introduced more formally than previously in the book. Probability theory underlies the powerful computational methodology known as simulation, which we introduced in Chapter 3. Simulation has many applications in probability and statistics and also in many other fields, such as engineering, chemistry, physics, and economics.

4.1   Basic Probability Calculations

The calculation of probabilities for random variables can often be simplified by tabulating the cumulative distribution function. Also, means and variances are easily calculated using component-wise column operations in Minitab. For example, suppose we have the probability distribution

x	1	2	3	4
probability	.1	.2	.3	.4

in columns C1 and C2, with the values in C1 and the probabilities in C2. The Calc ▶ Calculator command with the dialog box as in Display 4.1 computes the cumulative distribution function in C3 using Partial Sums.

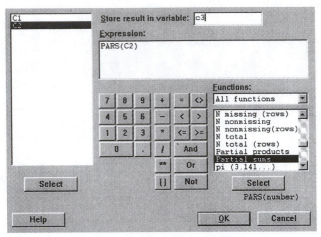

Display 4.1: Dialog box for computing partial sums of entries in C2 and placing these sums in C3.

Printing C1 and C3 gives

```
Row   C1    C3
1     1     0.1
2     2     0.3
3     3     0.6
4     4     1.0
```

in the Session window. We can also easily compute the mean and variance of this distribution. For example, the session commands

```
MTB > let c4=c1*c2
MTB > let c5=c1*c1*c2
MTB > let k1=sum(c4)
MTB > let k2=sum(c5)-k1*k1
MTB > print k1 k2
K1 3.00000
K2 1.00000
```

calculate the mean and variance and store these in K1 and K2, respectively. The mean is 3 and the variance is 1. Of course, we can also use Calc ▶ Calculator to do these calculations. In presenting more extensive computations, it is somewhat easier to list the appropriate session commands, as we will do subsequently. However, this is not to be interpreted as the required way to do these computations, as it is obvious that the menu commands can be used as well. Use whatever you find most convenient.

4.2 More on Sampling from Distributions

As we saw in II.3.2, Minitab includes algorithms for generating from many probability distributions using Calc ▶ Random Data. This menu command produces a drop-down list that includes the normal, binomial, Chi-square, F, t, uniform, and many other distributions that the text, and this manual, will discuss. Clicking on one of these names results in a dialog box with entries to be filled in further specifying the distribution and the size of the sample.

For example, we can generate from one particularly important class of probability distributions using Çalc ▶ Random Data ▶ Discrete. These probability distributions are concentrated on a finite number of values. To illustrate this, suppose we have the following values in C1 and C2.

Row	C1	C2
1	-1	0.3
2	2	0.2
3	3	0.4
4	10	0.1

Here, C1 contains the possible values of an outcome, and C2 contains the probabilities that each of these values is obtained, so, for example, $P(\{-1\}) = .3, P(\{2\}) = .2$, etc. The dialog box of Display 4.2 generates a sample of 50 from this discrete distribution and stores the sample in C3.

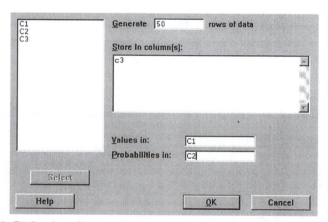

Display 4.2: Dialog box for generating a sample from a discrete distribution with values in C1 and probabilities in C2 and storing the sample in C3.

It is an interesting exercise to check that the algorithms Minitab is using are in fact producing samples appropriately. There are a variety of things one could check, but perhaps the simplest is to check that the long-run relative frequencies are correct. So in the example of this section, we want to make sure that, as we increase the size of the sample, the relative frequencies of $-1, 2, 3, 10$ in the sample are getting closer to .3, .2, .4, and .1, respectively. Note that it is not guaranteed that as we increase the sample size that the relative frequencies get closer monotonically to the corresponding probabilities, but inevitably this must be the case.

First, we generated a sample of size 100 from this distribution and stored the values in C3 as in Display 4.2. Next, we recorded a 1 in C4 whenever the corresponding entry in C3 was −1 and recorded a 0 in C4 otherwise. To do this, we used the Çalc ▶ Calculator command with dialog box as shown in Display 4.3.

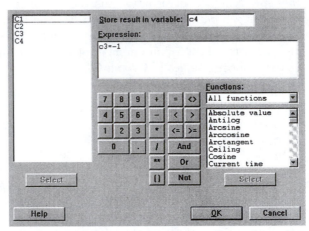

Display 4.3: Dialog box to record the incidence of a −1 in C3.

It is clear that the mean of C4 is the relative frequency of −1 in the sample. We calculated this mean using Çalc ▶ Çolumn Statistics, as discussed in I.10.3, which gave the output

```
Mean of C4 = 0.33000
```

in the Session window. Repeating this with a sample of size 1000, we obtained

```
Mean of C4 = 0.28100
```

which we can see is a bit closer to the true value of .3. Repeating this with a sample of size 10, 000 from this distribution, we obtained

```
Mean of C4 = 0.29300
```

which is closer still. It would appear that the relative frequency of −1 is indeed converging to .3.

We can generate a randomly chosen point from the line interval (a, b), where $a < b$, using Çalc ▶ Random Data ▶ Uniform. For example, the dialog box of Display 4.4 generates a sample of 1500 from the uniform distribution on the interval $(3.0, 6.3)$. With this distribution, the probability of any subinterval (c, d) of (a, b) is given by $(d − c) / (b − a)$, i.e., the length of (c, d) over the length of (a, b). Of course, we can estimate this probability by just counting the number of times the generated response falls in the interval (c, d) and dividing this by the total sample size. For example, using the outcomes from the dialog box of Display 4.3 and estimating the probability of the interval $(4, 5)$, we get the relative frequency 0.30867, which is close to the true value of $(5 − 4) / (6.3 − 3) = 0.30303$.

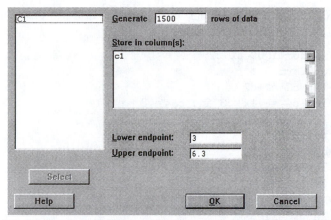

Display 4.4: Dialog box for generating a sample of 1500 from the uniform
distribution on the interval $(3.0, 6.3)$.

We can generalize this to generate from a point randomly chosen from a
rectangle $(a, b) \times (c, d)$, i.e., the set of all points (x, y) such that $a < x < b, c <
y < d$. If we want a sample of n from this distribution, we generate a sample
$x_1, \ldots, x_n$ from the uniform on (a, b) and also generate a sample $y_1, \ldots, y_n$ from
the uniform distribution on (c, d). Then $(x_1, y_1), \ldots, (x_n, y_n)$ is a sample of
n from the uniform distribution on $(a, b) \times (c, d)$. We can approximate the
probability of a random pair (x, y) falling in any subset $A \subset (a, b) \times (c, d)$ by
computing the relative frequency of A in the sample.

The **random** command is the session command for carrying out simulations
in Minitab. For example, the subcommand

uniform V_1 V_2

specifies the continuous uniform distribution on the interval (V_1, V_2); i.e., subintervals of the same length have the same probability of occurring. If we have
placed a discrete probability distribution in column E_2, on the values in column
E_1, the subcommand

discrete E_1 E_2

generates a sample from this distribution.

4.3 Simulation for Approximating Probabilities

As previously noted, simulation can be used to approximate probabilities. For
a variety of reasons, these simulations are most easily presented using session
commands but it is clear that we can replace each step by the appropriate menu
command.

For example, suppose we are asked to calculate

$$P(.1 \leq X_1 + X_2 \leq .3)$$

when X_1, X_2 are both independent and follow the uniform distribution on the interval $(0, 1)$. The session commands

```
MTB > random 1000 c1 c2;
SUBC> uniform 0 1.
MTB > let c3=c1+c2
MTB > let c4 = .1<=c3 and c3<=.3
MTB > let k1=sum(c4)/n(c4)
MTB > print k1
K1 0.0400000
MTB > let k2=sqrt(k1*(1-k1)/n(c4))
MTB > print k2
K2 0.00619677
MTB > let k3=k1-3*k2
MTB > let k4=k1+3*k2
MTB > print k3 k4
K3 0.0214097
K4 0.0585903
```

generate $N = 1000$ independent values of X_1, X_2 and place these values in C1 and C2, respectively, then calculate the sum $X_1 + X_2$ and put these values in C3. Using the comparison operators discussed in I.10.4, a 1 is recorded in C4 every time $.1 \le X_1 + X_2 \le .3$ is true and a 0 is recorded there otherwise. We then calculate the proportion of 1's in the sample as K1, and this is our estimate $\hat{p}$ of the probability. We will see later that a good measure of the accuracy of this estimate is the *standard error of the estimate*, which in this case is given by

$$\sqrt{\hat{p}(1-\hat{p})/N}$$

and this is computed in K2. Actually, we can feel fairly confident that the true value of the probability is in the interval

$$\hat{p} \pm 3\sqrt{\hat{p}(1-\hat{p})/N}$$

which in this case, equals the interval $(0.0214097, 0.0585903)$. So we know the true value of the probability with reasonable accuracy. As the simulation size N increases, the Law of Large Numbers says that $\hat{p}$ converges to the true value of the probability.

4.4   Simulation for Approximating Means

The means of distributions can also be approximated using simulations in Minitab. For example, suppose X_1, X_2 are both independent and follow the uniform distribution on the interval $(0, 1)$ and that we want to calculate the mean of $Y = 1/(1 + X_1 + X_2)$. We can approximate this in a simulation. The session commands

```
MTB > random 1000 c1 c2;
SUBC> uniform 0 1.
MTB > let c3=1/(1+c1+c2)
MTB > let k1=mean(c3)
MTB > let k2=stdev(c3)/sqrt(n(c3))
MTB > print k1 k2
K1 0.521532
K2 0.00375769
MTB > let k3=k1-3*k2
MTB > let k4=k1+3*k2
MTB > print k3 k4
K3 0.510259
K4 0.532805
```

generate $N = 1000$ independent values of X_1, X_2 and place these values in C1, C2, then calculate $Y = 1/(1 + X_1 + X_2)$ and put these values in C3. The mean of C3 is stored in K1, and this is our estimate of the mean value of Y. As a measure of how accurate this estimate is, we compute the standard error of the estimate, which is given by the standard deviation divided by the square root of the simulation sample size N. Again, we can feel fairly confident that the interval given by the estimate plus or minus 3 times the standard error of the estimate contains the true value of the mean. In this case, this interval is given by $(0.510259, 0.532805)$, and so we know this mean with reasonable accuracy. As the simulation size N increases, the Law of Large Numbers says that the approximation converges to the true value of the mean.

4.5   Exercises

When the data for an exercise come from an exercise in IPS, the IPS exercise number is given in parentheses ( ). All computations in these exercises are to be carried out using Minitab, and the exercises are designed to ensure that you have a reasonable understanding of the Minitab material in this chapter. Generally, you should be using Minitab to do all the computations and plotting required for the problems in IPS.

If your version of Minitab places restrictions such that the value of the simulation sample size N requested in these problems is not feasible, then substitute a more appropriate value. Be aware, however, that the accuracy of your results is dependent on how large N is.

1. Suppose we have the probability distribution

x	1	2	3	4	5
probability	.15	.05	.33	.37	.10

on the values 1, 2, 3, 4, and 5. Calculate the mean and variance of this distribution. Suppose that three independent outcomes (X_1, X_2, X_3) are generated from this distribution. Compute the probability that $1 < X_1 \leq 4, 2 \leq X_2$ and $3 < X_3 \leq 5$.

2. Suppose we have the probability distribution

x	1	2	3	4	5
probability	.15	.05	.33	.37	.10

on the values 1, 2, 3, 4, and 5. Using Minitab, verify that this is a probability distribution. Make a bar chart (probability histogram) of this distribution. Generate a sample of size 1000 from this distribution and plot a relative frequency histogram for the sample.

3. (4.23) Indicate how you would simulate the game of roulette using Minitab. Based on a simulation of $N = 1000$, estimate the probability of getting red and a multiple of 3.

4. A probability distribution is placed on the integers 1, 2, ..., 100, where the probability of integer i is c/i^2. Determine c so that this is a probability distribution. What is the 90th percentile? Generate a sample of 20 from the distribution.

5. Suppose an outcome is random on the square $(0, 1) \times (0, 1)$. Using simulation, approximate the probability that the first coordinate plus the second coordinate is less than .75 but greater than .25.

6. Generate a sample of 1000 from the uniform distribution on the unit disk $D = \left\{ (x, y) : x^2 + y^2 \leq 1 \right\}$.

7. The expression e^{-x} for $x > 0$ is the density curve for what is called the *Exponential* (1) distribution. Plot this density curve in the interval from 0 to 10 using an increment of .1. The Çalc ▶ Random Data ▶ Exponential command can be used to generate from this distribution by specifying the Mean as 1 in the ensuing dialog box. Generate a sample of 1000 from this distribution and estimate its mean. Approximate the probability that a value generated from this distribution is in the interval (1,2). The general *Exponential* (λ) has a density curve given by $\lambda^{-1} e^{-x/\lambda}$ for $x > 0$ and where $\lambda > 0$ is the mean. Repeat the simulation with mean $\lambda = 3$. Comment on the values of the estimated means.

8. Suppose you carry out a simulation to approximate the mean of a random variable X and you report the value 1.23 with a standard error of .025. If you are asked to approximate the mean of $Y = 3 + 5X$, do you have to carry out another simulation? If not, what is your approximation, and what is the standard error of this approximation?

9. Suppose that a random variable X follows an $N(3, 2.3)$ distribution. Subsequently, conditions change and no values smaller than -1 or bigger than 9.5 can occur, i.e., the distribution is conditioned to the interval $(-1, 9.5)$.

Generate a sample of 1000 from the truncated distribution, and use the sample to approximate its mean.

10. Suppose that X is a random variable and follows an $N(0, 1)$ distribution. Simulate $N = 1000$ values from the distribution of $Y = X^2$, and plot these values in a histogram with cutpoints 0, .5, 1, 1.5, ..., 15. Approximate the mean of this distribution. Generate Y directly from its distribution, which is known to be a *Chisquare*(1) distribution. In general, the *Chisquare*(k) distribution can be generated from via the command Calc ▶ Random Data ▶ Chi-Square, where k is specified as the Degrees of freedom in the dialog box. Plot the Y values in a histogram using the same cutpoints. Comment on the two histograms. Note that you can plot the density curve of these distributions using Calc ▶ Probability Distributions ▶ Chi-Square and evaluating the probability density at a range of points as we discussed in II.2 for the normal distribution.

11. If X_1 and X_2 are independent random variables with X_1 following a *Chisquare*(k_1) distribution and X_2 following a *Chisquare*(k_2) distribution, then it is known that $Y = X_1 + X_2$ follows a *Chisquare*($k_1 + k_2$) distribution. For $k_1 = 1$, $k_2 = 1$, verify this empirically by plotting histograms with cutpoints 0, .5, 1, 1.5, ..., 15, based on simulations of size $N = 1000$.

12. If X_1 and X_2 are independent random variables with X_1 following an $N(0, 1)$ distribution and X_2 following a *Chisquare*(k) distribution, then it is known that

$$Y = \frac{X_1}{\sqrt{X_2/k}}$$

follows a *Student*(k) distribution. The *Student*(k) distribution can be generated from using the command Calc ▶ Random Data ▶ t, where k is the Degrees of freedom and must be specified in the dialog box. For $k = 3$, verify this result empirically by plotting histograms with cutpoints -10, -9, ..., 9, 10, based on simulations of size $N = 1000$.

13. If X_1 and X_2 are independent random variables with X_1 following a *Chisquare*(k_1) distribution and X_2 following a *Chisquare*(k_2) distribution, then it is known that

$$Y = \frac{X_1/k_1}{X_2/k_2}$$

follows an $F(k_1, k_2)$ distribution. The $F(k_1, k_2)$ distribution can be generated from using the subcommand Calc ▶ Random Data ▶ F, where k_1 is the Numerator degrees of freedom and k_2 is the Denominator degrees of freedom, both of which must be specified in the dialog box. For $k_1 = 1$, $k_2 = 1$, verify this empirically by plotting histograms with cutpoints 0, .5, 1, 1.5, ..., 15, based on simulations of size $N = 1000$.

Chapter 5

Sampling Distributions

New Minitab command discussed in this chapter
Çalc ▶ Probability Distributions ▶ Binomial

Once data have been collected, they are analyzed using a variety of statistical techniques. Virtually, all of these involve computing *statistics* that measure some aspect of the data concerning questions we wish to answer. The answers determined by these statistics are subject to the uncertainty caused by the fact that we typically do not have the full population but only a sample from the population. As such, we have to be concerned with the variability in the answers when different samples are obtained. This leads to a concern with the *sampling distribution* of a statistic.

Sometimes, the sampling distribution of a statistic can be worked out exactly through various mathematical techniques, e.g., in Chapter 5 of IPS it is seen that the number of 1's in a sample of n from a *Bernoulli(p)* distribution is *Binomial(n,p)*. Often, however, this is not possible, and we must resort to approximations. One approximation technique is to use simulation. Sometimes, however, the statistics we are concerned with are averages, and, in such cases, we can typically approximate their sampling distribution via an appropriate normal distribution.

5.1   The Binomial Distribution

Suppose that $X_1, \ldots, X_n$ is a sample from the *Bernoulli(p)* distribution, i.e., $X_1, \ldots, X_n$ are independent realizations, where each X_i takes the value 1 or 0 with probabilities p and $1 - p$, respectively. Therandom variable $Y = X_1 + \cdots + X_n$ equals the number of 1's in the sample and follows, as discussed in IPS, a *Binomial(n,p)* distribution. Therefore, Y can take on any of the values $0, 1, \ldots, n$ with positive probability. In fact, an exact formula can be derived for these probabilities; namely

$$P(Y = k) = \binom{n}{k} p^k (1 - p)^{n-k}$$

is the probability that Y takes the value k for $0 \leq k \leq n$. When n and k are small, this formula could be used to evaluate this probability but it is almost always better to use software like Minitab to do it, and when these values are not small, it is necessary. Also, we can use Minitab to compute the $Binomial(n, p)$ cumulative probability distribution — the probability contents of intervals $(-\infty, x]$ and the inverse cumulative distribution — percentiles of the distribution.

For individual probabilities, we use the Çalc ▶ Probability Ðistributions ▶ Ƀinomial command. For example, suppose we have a $Binomial(30, .2)$ distribution and want to compute the probability $P(Y = 10)$. This command, with the dialog box as in Display 5.1, produces the output

```
Binomial with n = 30 and p = 0.200000
    x        P( X = x )
  10.00        0.0355
```

in the Session window, i.e., $P(Y = 10) = .0355$.

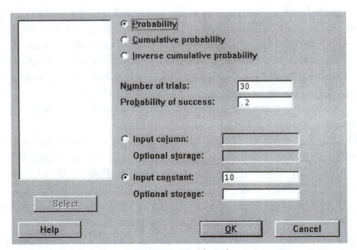

Display 5.1: Dialog box for $Binomial(n, p)$ probability calculations.

If we want to compute the probability of getting 10 or fewer successes, this is the probability of the interval $(-\infty, 10]$, and we can use the Çalc ▶ Probability Ðistributions ▶ Ƀinomial command with the dialog box as in Display 5.2. This produces the output

```
Binomial with n = 30 and p = 0.200000
    x        P( X <= x )
  10.00        0.9744
```

in the Session window, i.e., $P(Y \leq 10) = .9744$.

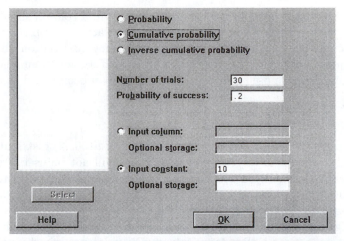

Display 5.2: Dialog box for computing cumulative probabilities for the
$Binomial(n, p)$ distribution.

Suppose we want to compute the first quartile of this distribution. The Çalc
▶ Probability Ðistributions ▶ Binomial command, with the dialog box as in
Display 5.3, produces the output

```
Binomial with n = 30 and p = 0.200000
  x     P( X <= x )     x       P( X <= x )
  3       0.1227        4         0.2552
```

in the Session window. This gives the values x that have cumulative probabilities
just smaller and just larger than the value requested. Recall that with a discrete
distribution, such as the $Binomial(n, p)$, we will not in general be able to obtain
an exact percentile.

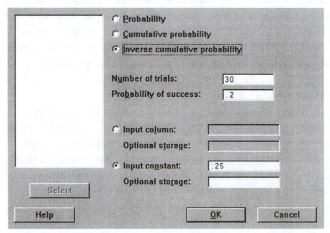

Display 5.3: Dialog box for computing percentiles of the $Binomial(n, p)$
distribution.

These commands can operate on all the values in a column simultaneously. This is very convenient if you should want to tabulate or graph the probability function, cumulative distribution function, or inverse distribution function.

The general syntax of the **pdf, cdf,** and **invcdf** session commands is given in II.1.3, and here we use them with the **binomial** subcommand as in

```
MTB > pdf 10;
SUBC> binomial 30 .2.
```

which outputs $P(Y = 10)$ when Y has the $Binomial(30, .2)$ distribution.

Actually, when n is very large even software will not be useful to compute these probabilities, and you will have to use normal approximations to binomial probabilities via the central limit theorem. The **pdf** and **cdf** commands with the **normal** subcommand can be used for this.

We might also want to simulate from the $Binomial(n, p)$ distribution. For this we use the Ca̲lc ▶ R̲andom Data ▶ B̲inomial command or the session command **random** with the **binomial** subcommand. For example,

```
MTB > random 10 c1;
SUBC> binomial 30 .2.
MTB > print c1
C1
 2 2 4 2 11 5 7 8 5 2
```

generates a sample of 10 from the $Binomial(30, .2)$ distribution.

5.2   Simulating Sampling Distributions

First, we consider an example where we know the exact sampling distribution. Suppose we flip a possibly biased coin n times and want to estimate the unknown probability p of getting a head. The natural estimate is $\hat{p}$ the proportion of heads in the sample. We would like to assess the sampling behavior of this statistic in a simulation. To do this, we choose a value for p, then generate N samples from the Bernoulli distribution of size n, for each of these compute $\hat{p}$, look at the empirical distribution of these N values, perhaps plotting a histogram as well. The larger N is the closer the empirical distribution and histogram will be to the true sampling distribution of $\hat{p}$.

Note that there are two sample sizes here: the sample size n of the original sample the statistic is based on, which is fixed, and the *simulation* sample size N, which we can control. This is characteristic of all simulations. Sometimes, using more advanced analytical techniques we can determine N so that the sampling distribution of the statistic is estimated with some prescribed accuracy. Some techniques for doing this are discussed in later chapters of IPS. Another method is to repeat the simulation a number of times, slowly increasing N until we see the results stabilize. This is sometimes the only way available, but caution should be shown as it is easy for simulation results to be very misleading if the final N is too small.

We illustrate a simulation to determine the sampling distribution of $\hat{p}$ when sampling from a *Bernoulli*(.75) distribution. For this, we use the commands Çalc ► Ŗandom Data ► Ḇernoulli, Çalc ► Rọw Statistics, and Ṣtat ► Ṭables ► Ṭa̱lly, with the dialog boxes given by Displays 5.4, 5.5, and 5.6, respectively, to produce the output

```
Summary Statistics for Discrete Variables
C11 CumPct
0.3   0.40
0.4   2.20
0.5   7.60
0.6  23.10
0.7  47.70
0.8  78.00
0.9  94.70
1.0 100.00
```

in the Session window. Here we have generated $N = 1000$ samples of size $n = 10$ from the *Bernoulli*(.75) distribution, i.e., we simulated the tossing of this coin 10,000 times, and we placed the results in the rows of columns C1–C10 using Çalc ► Ŗandom Data ► Ḇernoulli. The proportion of heads $\hat{p}$ in each sample is computed and placed in C11 using Çalc ► Rọw Statistics. Note that a mean of values equal to 0 or 1 is just the proportion of 1's in the sample. Finally, we used Ṣtat ► Ṭables ► Ṭa̱lly to compute the empirical distribution function of these 1000 values of $\hat{p}$. For example, this says 78% of these values were .8 or smaller and there were no instances smaller than .3.

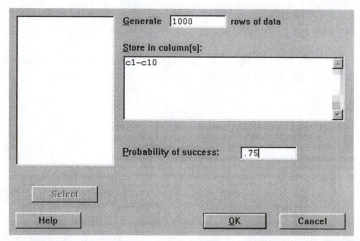

Display 5.4: Dialog box for generating 10 columns of 1000 *Bernoulli*(.75) values.

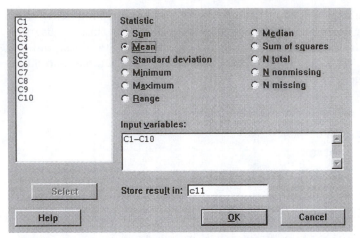

Display 5.5: Dialog box for computing the proportion of 1's in each of the 1000
samples of size 10.

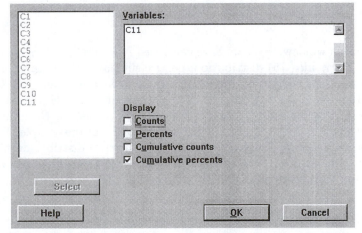

Display 5.6: Dialog box for computing the empirical distribution function of $\hat{p}$.

In Display 5.7, we have plotted a histogram of the 1000 values of $\hat{p}$. Based
on $N = 800$, the following empirical distribution was obtained:

```
C11   CumPct
 0.4    1.20
 0.5    7.20
 0.6   22.20
 0.7   47.80
 0.8   78.20
 0.9   95.00
 1.0  100.00
```

Because these values are reasonably close to those obtained with $N = 1000$, we
stopped at $N = 1000$.

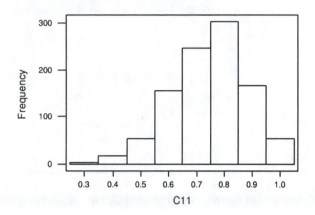

Display 5.7: Histogram of simulation of $N = 1000$ values of $\hat{p}$ based on a sample of size $n = 10$ from the Bernoulli(.75) distribution.

The corresponding session commands for this simulation are

```
MTB > random 1000 c1-c10;
SUBC> bernoulli .75.
MTB > rmean c1-c10 c11
MTB > tally c11;
SUBC> cumpcts.
```

and these might seem like an easier way to implement the simulation.

In Chapter 5 of IPS we saw that the sampling distribution of $\hat{p}$ can be determined exactly, i.e., there are formulas to determine this, and we can simulate directly from the sampling distribution, so this simulation can be made much more efficient. In effect, this entails using the Calc ▶ Random Data ▶ Binomial command with dialog box as in Display 5.8 and dividing each entry in C1 by 10. This generates $N = 1000$ values of $\hat{p}$ but uses a much smaller number of cells. Still, there are many statistics for which this kind of efficiency reduction is not available, and, to get some idea of what their sampling distribution is like, we must resort to the more brute force form of simulation of generating directly from the population distribution.

Sometimes, more sophisticated simulation techniques are needed to get an accurate assessment of a sampling distribution. Within Minitab, there are programming techniques, which we do not discuss in this manual, that can be applied in such cases. For example, it is clear that if our simulation required the generation of 10^6 cells (and this is not at all uncommon for some harder problems), the simulation approach we have described would not work within Minitab, as the worksheet would be too large.

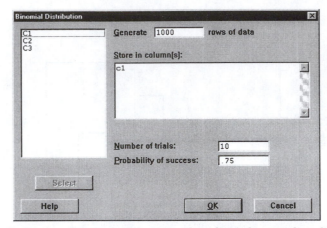

Display 5.8: Dialog box for generating 1000 values from the sampling distribution of $10\hat{p}$ using the *Binomial*(10, .75) distribution.

5.3   Exercises

When the data for an exercise come from an exercise in IPS, the IPS exercise number is given in parentheses ( ). All computations in these exercises are to be carried out using Minitab, and the exercises are designed to ensure that you have a reasonable understanding of the Minitab material in this chapter. Generally, you should be using Minitab to do all the computations and plotting required for the problems in IPS.

If your version of Minitab places restrictions such that the value of the simulation sample size N requested in these problems is not feasible, then substitute a more appropriate value. Be aware, however, that the accuracy of your results is dependent on how large N is.

1. Calculate all the probabilities for the *Binomial*(5, .4) distribution and the *Binomial*(5, .6) distribution. What relationship do you observe? Can you explain this and state a general rule?

2. Compute all the probabilities for a *Binomial*(5, .8) distribution and use these to directly calculate the mean and variance. Verify your answers using the formulas provided in IPS.

3. Compute and plot the probability and cumulative distribution functions of the *Binomial* (10, .2) and the *Binomial* (10, .5) distributions. Comment on the shapes of these distributions.

4. Generate 1000 samples of size 10 from the *Bernoulli*(.3) distribution. Compute the proportion of 1's in each sample and compute the proportion of samples having no 1's, one 1, two 1's, etc. Compute what these proportions would be in the longrun and compare.

5. Carry out a simulation study with $N = 1000$ of the sampling distribution of $\hat{p}$ for $n = 5, 10, 20$ and for $p = .5, .75, .95$. In particular, calculate the empirical distribution functions and plot the histograms. Comment on your findings.

6. Suppose that $X_1, X_2, X_3, \ldots$ are independent realizations from the *Bernoulli*(p) distribution, i.e., each X_i takes the value 1 or 0 with probabilities p and $1 - p$, respectively. If the random variable Y counts the number of tosses until we obtain the first head in a sequence of independent tosses $X_1, X_2, X_3, \ldots$, then Y has a *Geometric*(p) distribution. Minitab does not have built-in algorithms for computing the probability function, distribution function, inverse distribution function, and for generating from this distribution. The probability function for this distribution is given by

$$P(Y = y) = (1 - p)^{y-1} p$$

for $y = 1, 2, \ldots$. Plot the probability function for the *Geometric* (.5) distribution for the values $y = 1, \ldots, 10$. Do the same for the *Geometric* (.1) distribution. What do you notice?

7. Using methods for summing geometric sums, the cumulative distribution function of the *Geometric* (p) distribution (see Exercise II.5.6) is given by $P(Y \le y) = 1 - (1 - p)^y$. Plot the cumulative distribution function for the *Geometric* (.5) and *Geometric* (.1) distribution for the values $y = 1, \ldots, 10$. What do you notice?

8. To randomly generate from the *Geometric*(p) distribution (see Exercise II.5.6), we can repeatedly generate from a *Bernoulli*(p) and count how many times we did this until the first 1 appeared. A simple way to do this in Minitab is to generate N values from the *Bernoulli*(p) into a column. Count the number of entries until the first 1, count the number of subsequent entries until the next 1, etc. These counts are identically and independently distributed according to the *Geometric*(p) distribution. This is a very inefficient method when p is small and much better algorithms exist. Generate a sample of 10 from the *Geometric* (.5) distribution.

9. Carry out a simulation study, with $N = 2000$, of the sampling distribution of the sample standard deviation when sampling from the $N(0, 1)$ distribution, based on a sample of size $n = 5$. In particular, plot the histogram using cutpoints 0, 1.5, 2.0 2.5, 3.0 5.0. Repeat this for the sample coefficient of variation (sample standard deviation divided by the sample mean) using the cutpoints $-10, -9, \ldots, 0, \ldots, 9, 10$. Comment on the shapes of the histograms relative to a $N(0, 1)$ density curve.

10. Generate $N = 1000$ samples of size $n = 5$ from the $N(0, 1)$ distribution. Record a histogram for $\bar{x}$ using the cutpoints $-3, -2.5, -2, \ldots, 2.5, 3.0$. Generate a sample of size $N = 1000$ from the $N(0, 1/\sqrt{5})$ distribution. Plot the histogram using the same cutpoints and compare the histograms. What will happen to these histograms as we increase N?

11. Generate $N = 1000$ values of X_1, X_2, where X_1 follows a $N(3, 2)$ distribution and X_2 follows a $N(-1, 3)$ distribution. Compute $Y = X_1 - 2X_2$ for each of these pairs and plot a histogram for Y using the cutpoints $-20, -15, ..., 25, 30$. Generate a sample of $N = 1000$ from the appropriate distribution of Y and plot a histogram using the same cutpoints.

12. Plot the density curve for the *Exponential*(3) distribution (see Exercise II.4.7) between 0 and 15 with an increment of .1. Generate $N = 1000$ samples of size $n = 2$ from the *Exponential*(3) distribution and record the sample means. Standardize the sample of $\bar{x}$ using $\mu = 3$ and $\sigma = 3$. Plot a histogram of the standardized values using the cutpoints $-5, -4, ..., 4, 5$. Repeat this for $n = 5, 10$. Comment on the shapes of these histograms.

13. Plot the density of the uniform distribution on (0,1). Generate $N = 1000$ samples of size $n = 2$ from this distribution. Standardize the sample of $\bar{x}$ using $\mu = .5$ and $\sigma = \sqrt{1/12}$. Plot a histogram of the standardized values using the cutpoints $-5, -4, ..., 4, 5$. Repeat this for $n = 5, 10$. Comment on the shapes of these histograms.

14. The *Weibull*(β) has density curve given by $\beta x^{\beta-1} e^{-x^\beta}$ for $x > 0$, where $\beta > 0$ is a fixed constant. Plot the *Weibull*(2) density in the range 0 to 10 with an increment of .1 using the Çalc ▶ Probability Distributions ▶ Weibull, command. Generate a sample of $N = 1000$ from this distribution using the subcommand Çalc ▶ Random Data ▶ Weibull where β is the Shape parameter and the Scale parameter is 1. Plot a probability histogram and compare with the density curve.

Chapter 6

Introduction to Inference

In this chapter, the basic tools of statistical inference are discussed. There are a number of Minitab commands that aid in the computation of confidence intervals and for carrying out tests of significance.

6.1 z-Confidence Intervals

The command Stat ▶ Basic Statistics ▶ 1-Sample Z computes confidence intervals for the mean μ using a sample $x_1, \ldots, x_n$ from a distribution where we know the standard deviation σ. There are three situations when this is appropriate:

(1) We know that we are sampling from a normal distribution with unknown mean μ and known standard deviation σ, and thus

$$z = \frac{\bar{x} - \mu}{\sigma/\sqrt{n}}$$

is distributed $N(0, 1)$.

(2) We have a large sample from a distribution with unknown mean μ and known standard deviation σ, and the central limit theorem approximation to the distribution of $\bar{x}$ is appropriate, i.e., the distribution of

$$z = \frac{\bar{x} - \mu}{\sigma/\sqrt{n}}$$

is approximately distributed $N(0, 1)$.

105

(3) We have a large sample from a distribution with unknown mean μ and unknown standard deviation σ, and the sample size is large enough so that

$$z = \frac{\bar{x} - \mu}{s/\sqrt{n}}$$

is approximately $N(0, 1)$, where s is the sample standard deviation.

The confidence interval takes the form $\bar{x} \pm z^*\sigma/\sqrt{n}$, where s is substituted for σ in case (3), and z^* is determined from the $N(0,1)$ distribution by the confidence level desired, as described in IPS. Of course, situation (3) is probably the most realistic, but note that the confidence intervals constructed for (1) are exact, while those constructed under (2) and (3) are only approximate, and a larger sample size is required in (3) for the approximation to be reasonable than for (2).

Consider the sample given by 0.8403, 0.8363, 0.8447, which are stored in C1, and suppose that it makes sense to take $\sigma = .0068$. The command Stat ▶ Basic Statistics ▶ 1-Sample Z with the dialog boxes as in Displays 6.1 and 6.2 produces the output

```
Variable    N     Mean      StDev     SE Mean
   C1       3   0.84043    0.00420    0.00393
      99.0% CI
(0.83032, 0.85055)
```

in the Session window. This specifies the 99% confidence interval (0.83032, 0.85055) for μ. Note that in the dialog box of Display 6.1, we specify where the data resides in the Variables box, the value of σ in the Sigma box, and click on the Options button to bring up the dialog box in Display 6.2. In this dialog box we have specified the 99% confidence level in the Confidence level box.

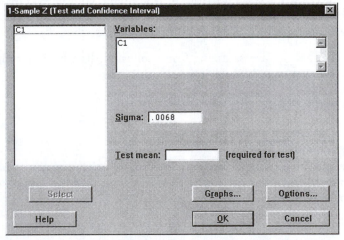

Display 6.1: First dialog box for producing the z-confidence interval for μ.

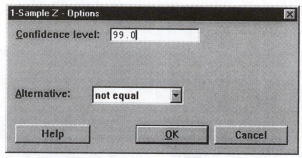

Display 6.2: Second dialog box for producing the z-confidence interval. Here we specify the confidence level.

The general syntax of the corresponding session command **zinterval** is

zinterval V_1 sigma $= V_2$ $E_1 \ldots E_m$

where V_1 is the confidence level and is any value between 1 and 99.99, V_2 is the assumed value of σ, and E_1, ..., E_m are columns of data. A $V_1\%$ confidence interval is produced for each column specified. If no value is specified for V_1, the default value is 95%.

6.2 z-Tests

The Stat ▶ Basic Statistics ▶ 1-Sample Z command is used when we want to test the hypothesis that the unknown mean μ equals a value μ_0 and one of the situations (1), (2), or (3) as discussed in II.10.1 is appropriate. The test is based on computing a P-value using the observed value of

$$z = \frac{\bar{x} - \mu_0}{\sigma/\sqrt{n}}$$

and the $N(0,1)$ distribution as described in IPS.

Suppose the sample $2.0, 0.4, 0.7, 2.0, -0.4, 2.2, -1.3, 1.2, 1.1, 2.3$ is stored in C1, and we are asked to test the null hypothesis $H_0 : \mu = 0$ against the alternative $H_a : \mu > 0$ and it makes sense to take $\sigma = 1$. The Stat ▶ Basic Statistics ▶ 1-Sample Z command together with the dialog boxes of Displays 6.3 and 6.4 produces the output

```
Variable      99.0% Lower Bound        Z        P
    C1               0.284           3.23    0.001
```

in the Session window. This specifies the P-value for this test as .001, and so we reject the null hypothesis in favor of the alternative. In the first dialog box, we specified where the data is located, the value of σ as before and that we want to test $H_0 : \mu = 0$ by 0 in the Test mean box. We brought up the second dialog box by clicking on the Options button. In the second dialog box, we specified that we want to test this null hypothesis against the alternative $H_a : \mu > 0$ by selecting "greater than" in Alternative box. The other choices are "not equal," which selects the alternative $H_a : \mu \neq 0$, and "less than," which selects the alternative $H_a : \mu < 0$.

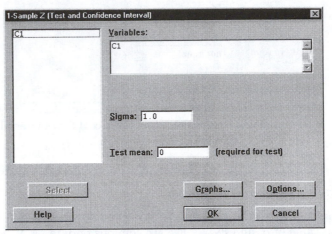

Display 6.3: First dialog box for testing a hypothesis concerning the mean using a z-test.

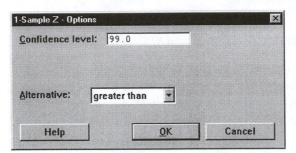

Display 6.4: Second dialog box for testing a hypothesis using the z-test.

The general syntax of the corresponding session command **ztest** is

ztest V_1 sigma $= V_2$ $E_1 \dots E_m$

where V_1 is the hypothesized value to be tested, V_2 is the assumed value of σ, and E_1, ..., E_m are columns of data. If no value is specified for V_1, the default is 0. A test of the hypothesis is carried out for each column. If no **alternative** subcommand is specified, a two-sided test is conducted, i.e., $H_0 : \mu = V_1$ against the alternative $H_a : \mu \neq V_1$. If the subcommand

    SUBC> alternative 1.

is used, a test of $H_0 : \mu = V_1$ against the alternative $H_a : \mu > V_1$ is conducted. If the subcommand

    SUBC> alternative -1.

is used, a test of $H_0 : \mu = V_1$ against the alternative $H_a : \mu < V_1$ is conducted.

6.3   Simulations for Confidence Intervals

When we are sampling from a $N(\mu, \sigma)$ distribution and know the value of σ, the confidence intervals constructed in II.6.1 are exact, i.e., in the long run a proportion 95% of the 95% confidence intervals constructed for an unknown mean μ will contain the true value of this quantity. Of course, any given confidence interval may or may not contain the true value of μ, and, in any finite number of such intervals so constructed, some proportion other than 95% will contain the true value of μ. As the number of intervals increases, however, the proportion covering will go to 95%.

We illustrate this via a simulation study based on computing 90% confidence intervals. The session commands

```
MTB > random 100 c1-c5;
SUBC> normal 1 2.
MTB > rmean c1-c5 c6
MTB > invcdf .95;
SUBC> normal 0 1.
Normal with mean = 0 and standard deviation = 1.00000
 P( X <= x) x
 0.9500 1.6449
MTB > let k1=1.6449*2/sqrt(5)
MTB > let c7=c6-k1
MTB > let c8=c6+k1
MTB > let c9=c7<1 and c8>1
MTB > mean c9
 Mean of C9 = 0.94000
MTB > set c10
DATA> 1:25
DATA> end
MTB > delete 26:100 c7 c8
MTB > mplot c7 versus c10 c8 versus c10;
SUBC> xstart=1 end=25;
SUBC> xincrement=1.
```

generate 100 random samples of size 5 from the $N(1, 2)$ distribution, place the means in C6, the lower end-point of a 90% confidence interval in C7, and the upper end-point in C8, and record whether or not a confidence interval covers the true value $\mu = 1$ by placing a 1 or 0 in C9, respectively. The mean of C9 is the proportion of intervals that cover, and this is 94%, which is 4% too high. Finally, we plotted the first 25 of these intervals in a plot shown in Figure 6.1. Drawing a solid horizontal line at 1 on the y-axis indicates that most of these intervals do indeed cover the true value $\mu = 1$.

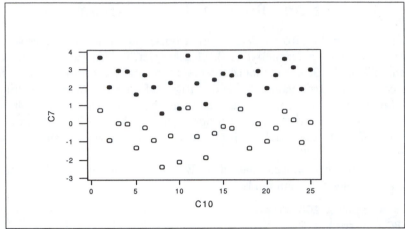

Figure 6.1: Plot of 90% confidence intervals for the mean when sampling from the $N(1, 2)$ distribution with $n = 5$. The lower end-point is open and the upper end-point is closed.

The simulation just carried out simply verifies a theoretical fact. On the other hand, when we are computing approximate confidence intervals — i.e., we are not sampling necessarily from a normal distribution — it is good to do some simulations from various distributions to see how much reliance we can place in the approximation at a given sample size. The true *coverage probability* of the interval, i.e., the long-run proportion of times that the interval covers the true mean, will not in general be equal to the nominal confidence level. Small deviations are not serious, but large ones are.

6.4   Simulations for Power Calculations

It is also useful to know in a given context how sensitive a particular test of significance is. By this we mean how likely it is that the test will lead us to reject the null hypothesis when the null hypothesis is false. This is measured by the concept of the *power* of a test. Typically, a level α is chosen for the P-value at which we would definitely reject the null hypothesis if the P-value is smaller than α. For example, $\alpha = .05$ is a common choice for this level. Suppose that we have chosen the level of .05 for the two-sided z-test and we want to evaluate the power of the test when the true value of the mean is $\mu = \mu_1$, i.e., evaluate the probability of getting a P-value smaller than .05 when the mean is μ_1. The two-sided z-test with level α rejects $H_0 : \mu = \mu_0$ whenever

$$P\left(|Z| > \left|\frac{\bar{x} - \mu_0}{\sigma/\sqrt{n}}\right|\right) \leq \alpha$$

where Z is a $N(0, 1)$ random variable. This is equivalent to saying that the null hypothesis is rejected whenever

$$\left| \frac{\bar{x} - \mu_0}{\sigma/\sqrt{n}} \right|$$

is greater than or equal to the $1 - \alpha/2$ percentile for the $N(0,1)$ distribution. For example, if $\alpha = .05$, then $1 - \alpha/2 = .975$ and this percentile can be obtained using the command Çalc ► Probability Ḏistributions ► Normal and the inverse distribution function, which gives the output

```
Normal with mean = 0 and standard deviation = 1.00000
  P( X <= x)         x
    0.9750       1.9600
```

in the Session window, i.e., the .975 percentile of the $N(0,1)$ distribution is 1.96. Denote this percentile by z^*. If $\mu = \mu_1$, then

$$\frac{\bar{x} - \mu_0}{\sigma/\sqrt{n}}$$

is a realized value from the distribution of $Y = \frac{\bar{X} - \mu_0}{\sigma/\sqrt{n}}$ when $\bar{X}$ is distributed $N(\mu_1, \sigma/\sqrt{n})$. Therefore, Y follows a $N(\frac{\mu_1 - \mu_0}{\sigma/\sqrt{n}}, 1)$ distribution. The power of the two-sided test at $\mu = \mu_1$ is

$$P(|Y| > z^*)$$

and this can be evaluated exactly using the command Çalc ► Probability Ḏistributions ► Ṇormal and the distribution function, after writing

$$P\left(|Y| > z^*\right) = P(Y > z^*) + P(Y < -z^*)$$
$$= P\left(Z > -\frac{(\mu_1 - \mu_0)}{\sigma/\sqrt{n}} + z^*\right) + P\left(Z < -\frac{(\mu_1 - \mu_0)}{\sigma/\sqrt{n}} - z^*\right)$$

with Z following an $N(0,1)$ distribution.

Alternatively, exact power calculations can be carried out under the assumption of sampling from a normal distribution using the Ṗower and Sample Size ► 1-Sample Ẕ command and filling in the dialog box appropriately. Also, the minimum sample size required to guarantee a given power at a prescribed difference $|\mu_1 - \mu_0|$ can be obtained using this command. For example, filling in the dialog box for this command as in Display 6.5 creates the output

```
Testing mean = null (versus not = null)
Calculating power for mean = null + difference
Alpha = 0.05 Sigma = 1.3
  Sample
Difference Size    Power
  0.1        10    0.0568
  0.2        10    0.0775
```

in the Session window. This gives the power for testing $H_0 : \mu = \mu_0$ versus $H_0 : \mu \neq \mu_0$ at $|\mu_1 - \mu_0| = .1$ and $|\mu_1 - \mu_0| = .2$ when $n = 10$, $\sigma = 1.3$, and $\alpha = .05$. These powers are given by .0568 and .0775, respectively. Clicking on the Options button allows you to choose other alternatives and specify other values of α in the Significance level box.

Power and Sample Size for 1-Sample Z

Specify values for any two of the following:

Sample sizes: `10`

Differences: `.1 .2`

Power values:

Sigma: `1.3`   Options...

Help   OK   Cancel

Display 6.5: Dialog box for calculating powers and minimum sample sizes.

If we had instead filled in Power values at .1 and .2 in the dialog box of Display 6.5, say as .8 and .9, and had left the Sample sizes box empty, we would have obtained the output

```
Testing mean = null (versus not = null)
Calculating power for mean = null + difference
Alpha = 0.05 Sigma = 1.3
  Sample Target Actual
  Difference    Size      Power      Power
        0.1     1327     0.8000     0.8002
        0.1     1776     0.9000     0.9000
        0.2      332     0.8000     0.8005
        0.2      444     0.9000     0.9000
```

in the Session window. This prescribes the minimum sample sizes $n = 1327$ and $n = 1776$ to obtain the powers .8 and .9, respectively, at the difference .1 and the sample sizes $n = 332$ and $n = 444$ to obtain the powers .8 and .9, respectively, at the difference .2.

This derivation of the power of the two-sided test depended on the sample coming from a normal distribution, as this leads to $\bar{X}$ having an exact normal distribution. In general, however, $\bar{X}$ will be only approximately normal, and so the normal calculation is not exact. To assess the effect of the nonnormality, however, we can often simulate sampling from a variety of distributions and estimate the probability $P(|Y| > z^*)$. For example, suppose that we want to test $H_0 : \mu = 0$ in a two-sided z-test based on a sample of 10, where we estimate σ by the sample standard deviation and we want to evaluate the power at 1. Let us further suppose that we are actually sampling from a uniform distribution on the interval $(-10, 12)$, which indeed has its mean at 1. The simulation given by the session commands

```
MTB > random 1000 c1-c10;
SUBC> uniform -10 12.
MTB > rmean c1-c10 c11
MTB > rstdev c1-c10 c12
MTB > let c13=absolute(c11/(c12/sqrt(10)))
MTB > let c14=c13>1.96
MTB > let k1=mean(c14)
MTB > let k2=sqrt(k1*(1-k1)/n(c14))
MTB > print k1 k2
K1 0.112000
K2 0.00997276
```

estimates the power to be .112, and the standard error of this estimate, as given in K2, is approximately .01. The application determines whether or not the assumption of a uniform distribution makes sense and whether or not this power is indicative of a sensitive test or not.

6.5   The Chi-Square Distribution

If Z is distributed according to the $N(0, 1)$ distribution, then $Y = Z^2$ is distributed according to the $Chisquare(1)$ distribution. If X_1 is distributed $Chisquare(k_1)$ independent of X_2 distributed $Chisquare(k_2)$, then $Y = X_1 + X_2$ is distributed according to the $Chisquare(k_1 + k_2)$ distribution. There are Minitab commands that assist in carrying out computations for the $Chisquare(k)$ distribution. Note that k is any positive value and is referred to as the *degrees of freedom*.

The values of the density curve for the $Chisquare(k)$ distribution can be obtained using the Calc ▶ Probability Distributions ▶ Chi-Square command, with k as the Degrees of freedom in the dialog box, or the session command **pdf** with the subcommand **chisquare.** For example, the command

```
MTB > pdf c1 c2;
SUBC> chisquare 4.
```

calculates the value of the $Chisquare(4)$ density curve at each value in C1 and stores these values in C2. This is useful for plotting the density curve. The Calc ▶ Probability Distributions ▶ Chi-Square command, or the session commands **cdf** and **invcdf,** can also be used to obtain values of the $Chisquare(k)$ cumulative distribution function and inverse distribution function, respectively. We use the Calc ▶ Random Data ▶ Chi-Square command, or the session command **random,** to obtain random samples from these distributions.

We will see applications of the chi-square distribution later in the book but we mention one here. In particular, if $x_1, \ldots, x_n$ is a sample from a $N(\mu, \sigma)$ distribution, then $(n-1)\, s^2/\sigma^2 = \sum_{i=1}^{n} (x_i - \bar{x})^2 /\sigma^2$ is known to follow a *Chisquare*$(n-1)$ distribution, and this fact is used as a basis for inference about σ (confidence intervals and tests of significance). Because of the nonrobustness of these inferences to small deviations from normality, these inferences are not recommended.

6.6   Exercises

When the data for an exercise come from an exercise in IPS, the IPS exercise number is given in parentheses ( ). All computations in these exercises are to be carried out using Minitab, and the exercises are designed to ensure that you have a reasonable understanding of the Minitab material in this chapter. Generally, you should be using Minitab to do all the computations and plotting required for the problems in IPS.

If your version of Minitab places restrictions such that the value of the simulation sample size N requested in these problems is not feasible, then substitute a more appropriate value. Be aware, however, that the accuracy of your results is dependent on how large N is.

1. (6.6) Use the Stat ▶ Basic Statistics ▶ 1- Sample Z command to compute 90%, 95%, and 99% confidence intervals for μ.

2. (6.49) Use the Stat ▶ Basic Statistics ▶ 1- Sample Z command to test the null hypothesis against the appropriate alternative. Evaluate the power of the test with level $\alpha = .05$ at $\mu = 225$.

3. Simulate $N = 1000$ samples of size 5 from the $N(1, 2)$ distribution, and calculate the proportion of .90 z-confidence intervals for the mean that cover the true value $\mu = 1$.

4. Simulate $N = 1000$ samples of size 10 from the uniform distribution on $(0,1)$, and calculate the proportion of .90 z-confidence intervals for the mean that cover the true value $\mu = .5$. Use $\sigma = 1/\sqrt{12}$.

5. Simulate $N = 1000$ samples of size 10 from the *Exponential*(1) distribution (see Exercise II.4.7), and calculate the proportion of .95 z-confidence intervals for the mean that cover the true value $\mu = 1$. Use $\sigma = 1$.

6. The density curve for the *Student*(1) distribution takes the form

$$\frac{1}{\pi} \frac{1}{1 + x^2}$$

for $-\infty < x < \infty$. This special case is called the *Cauchy* distribution. Plot this density curve in the range $(-20, 20)$ using an increment of .1. Simulate $N = 1000$ samples of size 5 from the *Student*(1) distribution (see Exercise

II.4.12), and calculate the proportion of .90 confidence intervals for the mean, using the sample standard deviation for σ, that cover the value $\mu = 0$. It is possible to obtain very bad approximations in this example because the central limit theorem does not apply to this distribution. In fact, it does not have a mean.

7. Suppose we are testing $H_0 : \mu = 3$ versus $H_0 : \mu \neq 3$ when we are sampling from a $N(\mu, \sigma)$ distribution with $\sigma = 2.1$ and the sample size is $n = 20$. If we use the critical value $\alpha = .01$, determine the power of this test at $\mu = 4$.

8. Suppose we are testing $H_0 : \mu = 3$ versus $H_0 : \mu > 3$ when we are sampling from a $N(\mu, \sigma)$ distribution with $\sigma = 2.1$. If we use the critical value $\alpha = .01$, determine the minimum sample size so that the power of this test at $\mu = 4$ is .99.

9. The uniform distribution on the interval (a, b) has mean $\mu = (a + b)/2$ and standard deviation $\sigma = \sqrt{(b-a)^2/12}$. Calculate the power at $\mu = 1$ of the two-sided z-test at level $\alpha = .95$ for testing $H_0 : \mu = 0$ when the sample size is $n = 10$, σ is the standard deviation of a uniform distribution on $(-10, 12)$, and we are sampling from a normal distribution.

10. Suppose that we are testing $H_0 : \mu = 0$ in a two-sided test based on a sample of 3. Approximate the power of the z-test at level $\alpha = .1$ at $\mu = 5$ when we are sampling from the distribution of $Y = 5 + W$, where W follows a *Student*(6) distribution (see Exercise II.4.12) and we use the sample standard deviation to estimate σ. Note that the mean of the distribution of Y is 5.

Chapter 7

Inference for Distributions

New Minitab commands discussed in this chapter

Çalc ▶ Probability Distributions ▶ F
Çalc ▶ Probability Distributions ▶ t
Çalc ▶ Random Data ▶ F
Çalc ▶ Random Data ▶ t
Power and Sample Size ▶ 1-Sample t
Power and Sample Size ▶ 2-Sample t
Stat ▶ Basic Statistics ▶ 1-Sample t
Stat ▶ Basic Statistics ▶ 2-Sample t
Stat ▶ Nonparametrics ▶ 1-Sample Sign

7.1 The Student Distribution

If Z is distributed $N(0,1)$ independent of X distributed $Chisquare(k)$ (see II.6.5), then $T = Z/\sqrt{X/k}$ is distributed according to the $Student(k)$ distribution. The value k is referred to as the *degrees of freedom* of the Student distribution. There are Minitab commands that assist in carrying out computations for this distribution.

The values of the density curve, distribution function, and inverse distribution function for the $Student(k)$ distribution can be obtained using the Çalc ▶ Probability Distributions ▶ t command with k as the Degrees of freedom. Alternatively, we can use the session commands **pdf, cdf,** and **invcdf** with the **student** subcommand. For example, the command

```
MTB > pdf c1 c2;
SUBC> student 4.
```

calculates the value of the $Student(4)$ density curve at each value in C1 and stores these values in C2. This is useful for plotting the density curve. To

117

generate from this distribution we use the command $\underline{C}$alc ▶ $\underline{R}$andom Data ▶ $\underline{t}$ again with k as the $\underline{D}$egrees of freedom or use the session command **random** with the **student** subcommand.

7.2 t-Confidence Intervals

When sampling from the $N(\mu, \sigma)$ distribution with μ and σ unknown, an exact $1 - \alpha$ confidence interval for μ based on the sample $x_1, \ldots, x_n$ is given by $\bar{x} \pm t^* s / \sqrt{n}$, where t^* is the $1 - \alpha/2$ percentile of the $Student(n-1)$ distribution. These intervals can be obtained using the $\underline{S}$tat ▶ $\underline{B}$asic Statistics ▶ $\underline{1}$- Sample t command. Suppose we place the data in Example 7.1 of IPS in column C1. This command, with the dialog box as in Display 7.1, produces the output

```
Variable    N    Mean    StDev    SE Mean      95.0 % CI
C1          8   22.50    7.19      2.54     (16.48,  28.52)
```

in the Session window. This computes a 95% confidence interval for μ as (16.48, 28.52). To change the confidence level, click on the $\underline{O}$ptions button and fill in the dialog box appropriately.

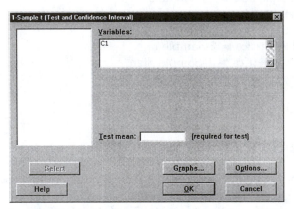

Display 7.1: Dialog box for producing t-confidence intervals.

Note that the $\underline{S}$tat ▶ $\underline{B}$asic Statistics ▶ $\underline{1}$- Sample t command can also be used to obtain confidence intervals for the difference of two means in a *matched pairs* design. For this, store the difference of the measurements in a column and apply $\underline{S}$tat ▶ $\underline{B}$asic Statistics ▶ $\underline{1}$- Sample t to that column as above.

The general syntax of the corresponding session command **tinterval** is

tinterval V $E_1 \ldots E_m$

where V is the confidence level and is any value between 1 and 99.99 and E_1, ..., E_m are columns of data. A V% confidence interval is produced for each column specified. If no value is specified for V, the default value is 95%.

7.3   *t*-Tests

The $\underline{\text{S}}$tat ▶ $\underline{\text{B}}$asic Statistics ▶ $\underline{1}$-Sample t command is used when we have
a sample $x_1, \ldots, x_n$ from a normal distribution with unknown mean μ and
standard deviation σ and we want to test the hypothesis that the unknown
mean equals a value μ_0. The test is based on computing a *P*-value using the
observed value of

$$t = \frac{\bar{x} - \mu_0}{s/\sqrt{n}}$$

and the *Student*$(n-1)$ distribution as described in IPS.

The command

```
MTB > ttest 40 c1
Test of mu = 40.00 vs mu not = 40.00
Variable   N    Mean    StDev   SE Mean    T       P
C1         8    22.50   7.19    2.54     -6.88   0.0002
```

calculates the *P*-value as .0002.

Suppose we have the data of Example 7.2 stored in column C1 and we want
to test the null hypothesis $H_0 : \mu = 40$ versus the alternative $H_a : \mu \neq 40$.
The $\underline{\text{S}}$tat ▶ $\underline{\text{B}}$asic Statistics ▶ $\underline{1}$-Sample t command with the dialog boxes as in
Displays 7.2 produces the output

```
Test of mu = 40.00 vs mu not = 40.00
Variable   N    Mean    StDev   SE Mean    T       P
C1         8    22.50   7.19    2.54     -6.88   0.0002
```

in the Session window. This calculates the *P*-value as .0002 and so we reject
H_0.

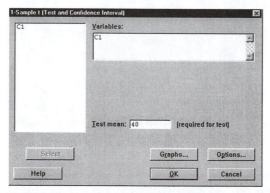

Display 7.2: First dialog box for a test of hypothesis using the *t*-statistic.

The general syntax of the corresponding session command **ttest** is

ttest V $E_1 \ldots E_m$

where V is the hypothesized value to be tested and $E_1, \ldots, E_m$ are columns of
data. If no value is specified for V, the default is 0. A test of the hypothesis

is carried out for each column. Also, the **alternative** subcommand is available and works just as with the **ztest** command.

Note that the Stat ▶ Basic Statistics ▶ 1-Sample t command can also be used to carry out t tests for the difference of two means in a matched pairs design. For this, store the difference of the measurements in a column and apply Stat ▶ Basic Statistics ▶ 1-Sample t to that column as above.

Exact power calculations can be carried under the assumption of sampling from a normal distribution using Power and Sample Size ▶ 1-Sample t and filling in the dialog box appropriately. Further, the minimum sample size required to guarantee a given power at a prescribed difference $|\mu_1 - \mu_0|$ and standard deviation σ can be obtained using this command. For example, using this command with the dialog box as in Display 7.3 we obtain the output

```
Testing mean = null (versus not = null)
Calculating power for mean = null + difference
Alpha = 0.05 Sigma = 3
  Sample
Difference    Size      Power
    2          5       0.2113
```

in the Session window. This gives the exact power of the two-sided t-test when $n = 5, |\mu_1 - \mu_0| = 2, \sigma = 3.0$, and $\alpha = .05$ as .2113. The Options button can be used to compute power for one-sided tests.

Display 7.3: Dialog box for determining power and minimum sample sizes when using the one sample t-test.

7.4   The Sign Test

As discussed in IPS, sometimes we cannot sensibly assume normality or transform to normality or make use of large samples so that there is a central limit theorem effect. In such a case, we attempt to use *distribution free* or *nonparametric* methods. The testing method based on the *sign test statistic* for the median is one of these.

For example, suppose we have the data for Example 7.1 in IPS stored in column C1. Then the S̲tat ▶ N̲onparametrics ▶ 1̲-Sample Sign command produces the dialog box given in Display 7.4.

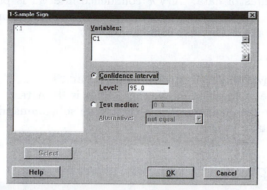

Display 7.4: Dialog box for the sign test and the sign confidence interval.

Here we have filled in the C̲onfidence interval button, and in the L̲evel box we have requested a .95 confidence interval for the median. The following output is obtained.

```
Sign confidence interval for median
                  Achieved
      N  Median Confidence    Confidence interval Position
C1    8   22.50  0.9297          (14.00, 31.00)        2
                 0.9500          (13.81, 31.00)       NLI
                 0.9922          (11.00, 31.00)        1
```

As the distribution of the sign statistic is discrete, in general the exact confidence cannot be attained so Minitab records the confidence intervals with confidence level just smaller and just greater than the confidence level requested and records a middle interval by interpolation. If instead we fill in the T̲est median button and enter 40 for the null hypothesis with the A̲lternative not equal we obtain the output

```
Sign test of median = 40.00 versus not = 40.00
      N   Below   Equal   Above      P    Median
C1    8     8       0       0     0.0078   22.50
```

which gives the *P*-value as .0078 for testing that the median of the population distribution equals 40 against the two-sided alternative that it doesn't equal this value. Also, the sample median of 22.5 is recorded.

Note that the S̲tat ▶ N̲onparametrics ▶ 1̲-Sample Sign command can also be used to construct confidence intervals and carry out tests for the median of a difference in a matched pairs design. For this, store the difference of the measurements in a column and apply the command to that column.

The corresponding session commands are **sinterval,** for the sign confidence interval and **stest,** for the sign test. The general syntax of the **sinterval** command is

sinterval V $E_1 \ldots E_m$

where V is the confidence level and is any value between 1 and 99.99 and E_1, ..., E_m are columns of data. A V% confidence interval is produced for each column specified. If no value is specified for V, then the default value is 95%. The general syntax of the **stest** command is

stest V $E_1 \ldots E_m$

where V is the hypothesized value to be tested and E_1, ..., E_m are columns of data. If no value is specified for V, the default is 0. A test of the hypothesis is carried out for each column. The **alternative** subcommand is also available for one-sided tests.

7.5   Comparing Two Samples

If we have independent samples $x_{11}, \ldots x_{1n_1}$ from the $N(\mu_1, \sigma_1)$ distribution and $x_{12}, \ldots x_{1n_2}$ from the $N(\mu_2, \sigma_2)$ distribution, where σ_1 and σ_2 are known, we can base inferences about the difference of the means $\mu_1 - \mu_2$ on the z-statistic given by

$$z = \frac{\bar{x}_1 - \bar{x}_2 - (\mu_1 - \mu_2)}{\sqrt{\frac{\sigma_1^2}{n_1} + \frac{\sigma_2^2}{n_2}}}.$$

Under these assumptions, z has an $N(0, 1)$ distribution. Therefore, a $1 - \alpha$ confidence interval for $\mu_1 - \mu_2$ is given by

$$\bar{x}_1 - \bar{x}_2 \pm \sqrt{\frac{\sigma_1^2}{n_1} + \frac{\sigma_2^2}{n_2}} z^*$$

where z^* is the $1 - \alpha/2$ percentile of the $N(0, 1)$ distribution. Further, we can test $H_0 : \mu = \mu_0$ against the alternative $H_a : \mu \neq \mu_0$ by computing the P-value $P(|Z| > |z_0|) = 2P(Z > z_0)$, where Z is distributed $N(0, 1)$ and z_0 is the observed value of the z-statistic. These inferences are also appropriate without normality, provided n_1 and n_2 are large and we have reasonable values for σ_1 and σ_2. These inferences are easily carried out using Minitab commands we have already discussed.

In general, however, we will not have available suitable values of σ_1 and σ_2 or large samples and will have to use the two-sample analogs of the single-sample t procedures just discussed. This is acceptable, provided, of course, that we have checked for and agreed that it is reasonable to assume that both samples are from normal distributions. These procedures are based on the two-sample t-statistic given by

$$t = \frac{\bar{x}_1 - \bar{x}_2 - (\mu_1 - \mu_2)}{\sqrt{\frac{s_1^2}{n_1} + \frac{s_2^2}{n_2}}}$$

where we have replaced the population standard deviations by their sample estimates. The exact distribution of this statistic does not have a convenient form, but, of course, we can always simulate its distribution. Actually, it is typical to use an approximation to the distribution of this statistic based on a Student distribution. See the discussion in IPS on this, and use Help to get more details.

The Stat ▶ Basic Statistics ▶ 2-Sample t command is available for computing inference procedures based on t. For example, suppose that we have the data for Example 7.20 of IPS in a worksheet with the Placebo sample in C1 and the Calcium sample in C2. The Stat ▶ Basic Statistics ▶ 2-Sample t command with the dialog box as in Display 7.5

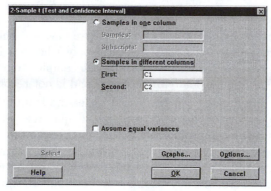

Display 7.5: Dialog box for two sample problems based on the two-sample t-statistic.

produces the output

```
Two-sample T for C1 vs C2
 N Mean StDev SE Mean
C1 10 5.00 8.74 2.8
C2 11 -0.64 5.87 1.8
Difference = mu C1 - mu C2
Estimate for difference:  5.64
95% CI for difference:  (-1.36, 12.63)
T-Test of difference = 0 (vs not =):   T-Value = 1.72
P-Value = 0.107 DF = 15
```

in the Session window. This gives a 95% confidence interval for the difference in the means $\mu_1 - \mu_2$ as $(-1.36, 12.63)$ and calculates the P-value .107 for the test of $H_0 : \mu_1 - \mu_2 = 0$ versus the alternative $H_a : \mu_1 - \mu_2 \neq 0$. In this case, we do not reject H_0. Notice we have selected the Samples in different columns radio button, as this is how we have stored our data. Alternatively, we can store all the actual measurements in a single column with a second column providing an index of the sample to which the observation belongs. Clicking on the Options button of the dialog box of Display 7.5 produces the dialog box of Display 7.6. From this, we see that we can prescribe a different value for the confidence level, the difference between the means that we wish to test for, and the type of hypothesis.

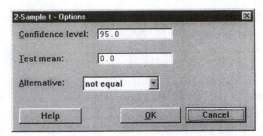

Display 7.6: Dialog box obtained when we click on the Options button in the dialog
box of the Stat ▶Basic Statistics ▶2-Sample t command.

Notice that in the dialog box of Display 7.5 we have left the box unchecked
Assume equal variances. This box is checked only when we feel that we can
assume that $\sigma_1 = \sigma_2 = \sigma$ and want to pool both samples together to estimate
the common σ. Pooling is usually unnecessary and is not recommended.

Exact power calculations can be carried under the assumption of sampling
from a normal distribution using Power and Sample Size ▶ 2-Sample t and
filling in the dialog box appropriately, although this requires the assumption
of a common population standard deviation σ. Further, the minimum sample
size required to guarantee a given power at a prescribed difference $|\mu_1 - \mu_2|$, and
assuming a common standard deviation σ, can be obtained using this command.
This command works the same as the one sample case.

There are two corresponding session commands — **twosample** and **twot**.
Each of these commands computes confidence intervals for the difference of the
means and computes P-values for tests of significance concerning the difference
of means. The only difference between these commands is that with **twosample**
the two samples are in individual columns, while with **twot** the samples are in a
single column with subscripts indicating group membership in a second column.
The general syntax of the **twosample** command is

    twosample V E_1 E_2

where V is the confidence level and is any value between 1 and 99.99 and E_1,
E_2 are columns of data containing the two samples. The general syntax of the
twot command is

    twot V E_1 E_2

where V is the confidence level and is any value between 1 and 99.99 and E_1,
E_2 are columns of data with E_1 containing the samples and E_2 containing the
subscripts.

The **alternative** subcommand is available with both **twosample** and **twot**
if we wish to conduct one-sided tests. Also, the subcommand **pooled** is available
if we feel we can assume that $\sigma_1 = \sigma_2 = \sigma$ and want to pool both samples
together to estimate the common σ.

7.6   The F-Distribution

If X_1 is distributed $Chisquare(k_1)$ independent of X_2 distributed $Chisquare(k_2)$, then

$$F = \frac{X_1/k_1}{X_2/k_2}$$

is distributed according to the $F(k_1, k_2)$-distribution. The value k_1 is called the *numerator degrees of freedom* and the value k_2 is called the *denominator degrees of freedom*. There are Minitab commands that assist in carrying out computations for this distribution.

The values of the density curve for the $F(k_1, k_2)$ distribution can be obtained using the Çalc ▶ Probability Distributions ▶ F command, with k_1 specified as the Numerator degrees of freedom and k_2 specified as the Denominator degrees of freedom in the dialog box. For example, this command with the dialog box as in Display 7.7 produces the output

```
     x            P( X <= x )
  5.0000            0.8288
```

in the Session window. This calculates the value of the $F(3, 2)$ distribution function at 5 as .8288. Alternatively, you can use the session commands **pdf**, **cdf**, and **invcdf** with the **F** subcommand. The Çalc ▶ Random Data ▶ F command and the session command **random** with the **F** subcommand can be used to obtain random samples from the $F(k_1, k_2)$ distribution.

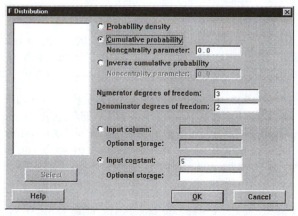

Display 7.7: Dialog box for probability calculations for the $F(k_1, k_2)$ distribution.

There are a number of applications of the F-distribution. In particular, if $x_{11}, \ldots x_{1n_1}$ is a sample from the $N(\mu_1, \sigma_1)$ distribution and $x_{12}, \ldots x_{1n_2}$ a sample from the $N(\mu_2, \sigma_2)$ distribution, then

$$F = \frac{s_1^2/\sigma_1^2}{s_2^2/\sigma_2^2}$$

is known to follow an $F(n_1 - 1, n_2 - 1)$. This fact is used as a basis for inference about the ratio σ_1/σ_2, i.e., confidence intervals and tests of significance and, in particular, testing for equality of variances between the samples. Because of the nonrobustness of these inferences to small deviations from normality, these inferences are not recommended.

7.7 Exercises

When the data for an exercise come from an exercise in IPS, the IPS exercise number is given in parentheses ( ). All computations in these exercises are to be carried out using Minitab, and the exercises are designed to ensure that you have a reasonable understanding of the Minitab material in this chapter. Generally, you should be using Minitab to do all the computations and plotting required for the problems in IPS.

If your version of Minitab places restrictions such that the value of the simulation sample size N requested in these problems is not feasible, then substitute a more appropriate value. Be aware, however, that the accuracy of your results is dependent on how large N is.

1. Plot the $Student(k)$ density curve for $k = 1, 2, 10, 30$ and the $N(0, 1)$ density curve on the interval $(-10, 10)$ using an increment of .1 and compare the plots.

2. Make a table of the values of the cumulative distribution function of the $Student(k)$ distribution for $k = 1, 2, 10, 30$ and the $N(0, 1)$ distribution at the points $-10, -5, -3, -1, 0, 1, 3, 5, 10$. Comment on the values.

3. Make a table of the values of the inverse cumulative distribution function of the $Student(k)$ distribution for $k = 1, 2, 10, 30$ and the $N(0, 1)$ distribution at the points $.0001, .001, .01, .1, .25, .5$. Comment on the values.

4. Simulate $N = 1000$ values from Z distributed $N(0, 1)$ and X distributed $Chisquare(3)$ and plot a histogram of $T = Z/\sqrt{X/3}$ using the cutpoints $-10, -9, \ldots, 9, 10$. Generate a sample of $N = 1000$ values directly from the $Student(3)$ distribution, plot a histogram with the same cutpoints, and compare the two histograms.

5. Carry out a simulation with $N = 1000$ to verify that the 95% confidence interval based on the t-statistic covers the true value of the mean 95% of the time when taking samples of size 5 from the $N(4, 2)$ distribution.

6. Generate a sample of 50 from the $N(10, 2)$ distribution. Compare the 95% confidence intervals obtained via the S̲tat ▶ B̲asic Statistics ▶ 1̲-Sample t and S̲tat ▶ B̲asic Statistics ▶ 1- Sample Z̲ commands using the sample standard deviation as an estimate of σ.

7. Calculate the power of the t-test at $\mu_1 = 1, \sigma_1 = 2$ for testing $H_0 : \mu = 0$ versus the alternative $H_a : \mu \neq 0$ at level $\alpha = .05$, based on a sample of 5 from the normal distribution.

8. Simulate the power of the two sample t-test at $\mu_1 = 1, \sigma_1 = 2, \mu_2 = 2, \sigma_1 = 3$ for testing $H_0 : \mu_1 - \mu_2 = 0$ versus the alternative $H_a : \mu_1 - \mu_2 \neq 0$ at level $\alpha = .05$, based on a sample of 5 from the $N(\mu_1, \sigma_1)$ distribution and a sample of size 8 from the $N(\mu_2, \sigma_2)$ distribution. Use the conservative rule when choosing the degrees of freedom for the approximate test, i.e., the smaller of $n_1 - 1$ and $n_2 - 1$.

9. If Z is distributed $N(\mu, 1)$ and X is distributed $Chisquare(k)$ independent of Z, then

$$Y = \frac{Z}{\sqrt{X/k}}$$

is distributed according to a *noncentral Student(k)* distribution with non-centrality μ. Simulate samples of $N = 1000$ from this distribution with $k = 5$ and $\mu = 0, 1, 5, 10$. Plot the samples in histograms with cutpoints $-20, -19, \ldots, 19, 20$ and compare these plots.

10. If X_1 is distributed $Chisquare(k_1)$ independently of X_2, which is distributed $N(\delta, 1)$, then the random variable $Y = X_1 + X_2^2$ is distributed according to a *noncentral Chisquare(k + 1)* distribution with noncentrality $\lambda = \delta^2$. Generate samples of $n = 1000$ from this distribution with $k = 2$ and $\lambda = 0, 1, 5, 10$. Plot histograms of these samples with the cut-points $0, 1, \ldots, 200$. Comment on the appearance of these histograms.

11. If X_1 is distributed *noncentral Chisquare(k_1)* with non-centrality λ independently of X_2, which is distributed $Chisquare(k_2)$, then the random variable

$$Y = \frac{X_1/k_1}{X_2/k_2}$$

is distributed according to a *noncentral F(k_1, k_2)* distribution with non-centrality λ. Generate samples of $n = 1000$ from this distribution with $k_1 = 2, k_2 = 3$, and $\lambda = 0, 1, 5, 10$. Plot histograms of these samples with the cut-points $0, 1, \ldots, 200$. Comment on the appearance of these histograms.

Chapter 8

Inference for Proportions

New Minitab commands discussed in this chapter

Power and Sample Size ▶ 1 Proportion
Power and Sample Size ▶ 2 Proportions
Stat ▶ Basic Statistics ▶ 1 Proportion
Stat ▶ Basic Statistics ▶ 2 Proportions

This chapter is concerned with inference methods for a proportion p and for the comparison of two proportions p_1 and p_2. Proportions arise from measuring a binary-valued categorical variable on population elements, such as gender in human populations. For example, p might be the proportion of females in a given population, or we might want to compare the proportion p_1 of females in population 1 with the proportion p_2 of females in population 2. The need for inference arises as we base our conclusions about the values of these proportions on samples from the populations rather than measuring every element in the population. For convenience, we will denote the values assumed by the binary categorical variables as 1 and 0, where 1 indicates the presence of a characteristic and 0 indicates its absence.

8.1 Inference for a Single Proportion

Suppose that $x_1, \ldots, x_n$ is a sample from a population where the variable is measuring the presence or absence of some trait by a 1 or 0, respectively. Let $\hat{p}$ be the proportion of 1's in the sample. This is the estimate of the true proportion p. For example, the sample could arise from coin tossing, where 1 denotes heads and 0 tails and $\hat{p}$ is the proportion of heads, while p is the probability of heads. If the population we are sampling from is finite, then, strictly speaking, the sample elements are not independent. But if the population size is large relative to the sample size n, then independence is a reasonable approximation, and this is necessary for the methods of this chapter. So we will consider $x_1, \ldots, x_n$ as a sample from the *Bernoulli(p)* distribution.

The standard error of the estimate $\hat{p}$ is $\sqrt{\hat{p}(1-\hat{p})/n}$, and because $\hat{p}$ is an average, the central limit theorem gives that

$$z = \frac{\hat{p} - p}{\sqrt{\frac{\hat{p}(1-\hat{p})}{n}}}$$

is approximately $N(0,1)$ for large n. This leads to the approximate $1 - \alpha$ confidence interval given by $\hat{p} \pm \sqrt{\hat{p}(1-\hat{p})/n} z^*$, where z^* is the $1 - \alpha/2$ percentile of the $N(01)$ distribution. To test a null hypothesis $H_0 : p = p_0$, we make use of the fact that under the null hypothesis the statistic

$$z = \frac{\hat{p} - p_0}{\sqrt{\frac{p_0(1-p_0)}{n}}}$$

is approximately $N(0,1)$. To test $H_0 : p = p_0$ versus $H_a : p \neq p_0$, we compute $P(|Z| > |z|) = 2P(Z > |z|)$, where Z is distributed $N(0,1)$.

For example, in Example 8.2 of IPS the probability of heads was estimated by Count Buffon as $\hat{p} = .5069 = 2048/4040$ on the basis of a sample of $n = 4040$ tosses. The command $\underline{S}$tat ▶ $\underline{B}$asic Statistics ▶ 1 $\underline{P}$roportion with the dialog box as in Display 8.1

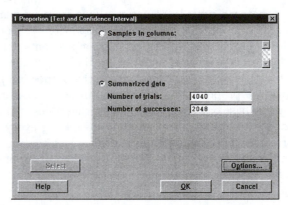

Display 8.1: Dialog box for inferences about single proportions.

produces the output

```
    Test of p = 0.5 vs p not = 0.5
                                                          Exact
    Sample      X     N    Sample p        95.0% CI      P-Value
       1     2048  4040    0.506931  (0.491390, 0.522461)  0.387
```

in the Session window. This computes the 90% confidence interval for p to be $(0.491390, 0.522461)$ and also computes the *P*-value for testing $H_0 : p = 0$ versus $H_a : p \neq 0$ as .387 so that we would not reject the null hypothesis. If we wish to use other confidence levels or test other hypotheses, then these options are all available using the Options button on this dialog box.

Note that the estimate and confidence intervals recorded by the software are not those based on the *Wilson estimate* discussed in IPS. To obtain the Wilson estimate and the associated confidence interval, we must add four data values to the data set — two heads (or successes) and two tails (or failures). So in this case, implementing the above command with the number of trials equal to 4044 and the number of successes equal to 2050 will produce the inferences based on the Wilson estimate.

Power calculations and minimum sample sizes to achieve a prescribed power can be obtained using Power and Sample Size ▶ 1 _Proportion. For example, suppose we want to compute the power of the test for $H_0 : p = .5$ versus $H_a : p \neq .5$ at level $\alpha = .05$ at $n = 10, p = .4$. This command, with the dialog box as in Display 8.2,

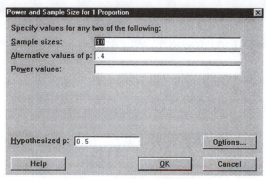

Display 8.2: Dialog box for power calculations for test of a single proportion.

produces the output

```
Testing proportion = 0.5 (versus not = 0.5)
Alpha = 0.05
     Alternative     Sample
     Proportion       Size       Power
       0.400000         10       0.0918
```

which calculates this power as .0918. So the test is not very powerful. By contrast, at $n = 100, p = .4$ the power is .51633.

8.2   Inference for Two Proportions

Suppose that $x_{11}, \ldots, x_{n_1 1}$ is a sample from population 1 and $x_{12}, \ldots, x_{n_2 2}$ is a sample from population 2, where the variable is measuring the presence or absence of some trait by a 1 or 0, respectively. We assume then that we have a sample of n_1 from the $Bernoulli(p_1)$ distribution and a sample of n_2 from the $Bernoulli(p_2)$ distribution. Suppose that we want to make inferences about the difference in the proportions $p_1 - p_2$. Let $\hat{p}_i$ be the proportion of 1's in the ith sample.

The central limit theorem gives that

$$z = \frac{\hat{p}_1 - \hat{p}_2 - (p_1 - p_2)}{\sqrt{\frac{\hat{p}_1(1-\hat{p}_1)}{n_1} + \frac{\hat{p}_2(1-\hat{p}_2)}{n_2}}}$$

is approximately $N(0,1)$ for large n_1 and n_2. This leads to the approximate $1 - \alpha$ confidence interval given by

$$\hat{p}_1 - \hat{p}_2 \pm \sqrt{\frac{\hat{p}_1(1-\hat{p}_1)}{n_1} + \frac{\hat{p}_2(1-\hat{p}_2)}{n_2}}\, z^*$$

where z^* is the $1 - \alpha/2$ percentile of the $N(0,1)$ distribution. As indicated in IPS, the Wilson estimate and its corresponding confidence interval are obtained by adding four data values to the data set — one success and one failure to each sample — so that the *ith* sample size becomes $n_i + 2$ and the *ith* sample estimate becomes $(n_i\hat{p}_i + 1)/(n_i + 2)$. The above formula for the confidence interval applied with these changes then gives the interval based on the Wilson estimates.

To test a null hypothesis $H_0 : p_1 = p_2$ we use the fact that under the null hypothesis the statistic

$$z = \frac{\hat{p}_1 - \hat{p}_2}{\sqrt{\hat{p}(1-\hat{p})\left(\frac{1}{n_1} + \frac{1}{n_2}\right)}}$$

is approximately $N(0,1)$ for large n_1 and n_2, where

$$\hat{p} = \frac{n_1\hat{p}_1 + n_2\hat{p}_2}{n_1 + n_2}$$

is the estimate of the common value of the proportion when the null hypothesis is true. To test $H_0 : p_1 = p_2$ versus $H_a : p_1 \neq p_2$ we compute $P(|Z| > |z|) = 2P(Z > |z|)$ where Z is distributed $N(0,1)$.

For example, in Example 8.9 of IPS suppose that we want to test $H_0 : p_1 = p_2$ versus $H_a : p_1 \neq p_2$ where $n_1 = 61, \hat{p}_1 = .803 = 49/61, n_2 = 62,$ $\hat{p}_2 = .613 = 38/62$. The command $\underline{S}$tat ▶ $\underline{B}$asic Statistics ▶ 2 P$\underline{r}$oportions with the dialog box as in Display 8.3 produces the output

```
    Sample    X      N      Sample p
      1       49     61     0.803279
      2       38     62     0.612903
Estimate for p(1) - p(2):  0.190375
95% CI for p(1) - p(2):  (0.0333680, 0.347383)
Test for p(1) - p(2) = 0 (vs not = 0):  Z = 2.38
P-Value = 0.017
```

in the Session window. The *P*-value is .017, so we would definitely reject. A 95% confidence interval for the difference $p_1 - p_2$ is given by

(0.0333680,0.347383). If other tests or confidence intervals are required, then these are available via the Options button. The Wilson estimates and associated confidence interval are obtained from the software by modifying the data as indicated above.

Power calculations and minimum sample sizes to achieve a prescribed power can be obtained using Power and Sample Size ▶ 2 Proportions.

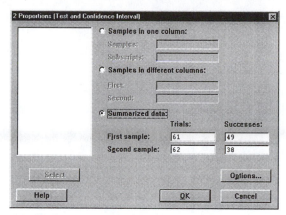

Display 8.3: Dialog box for inferences comparing two proportions.

8.3 Exercises

When the data for an exercise come from an exercise in IPS, the IPS exercise number is given in parentheses ( ). All computations in these exercises are to be carried out using Minitab, and the exercises are designed to ensure that you have a reasonable understanding of the Minitab material in this chapter. Generally, you should be using Minitab to do all the computations and plotting required for the problems in IPS.

Don't forget to quote standard errors for any approximate probabilities you quote in the following problems.

1. Carry out a simulation with the $Binomial(40, .3)$ distribution to assess the coverage of the 95% confidence interval for a single proportion.

2. The accuracy of a confidence interval procedure can be assessed by computing *probabilities of covering false values*. Approximate the probabilities of covering the values .1, .2, ..., .9 for the 95% confidence interval for a single proportion when sampling from the $Binomial(20, .5)$ distribution.

3. Calculate the power of the two-sided test for testing $H_0 : p = .5$ at level $\alpha = .05$ at the points $n = 100, p = .1, ..., 9$ and plot the power curve.

4. Carry out a simulation with the $Binomial(40, .3)$ and the $Binomial(50, .4)$ distribution to assess the coverage of the 95% confidence interval for a difference of proportions.

5. Calculate the power of the two-sided test for testing $H_0 : p_1 = p_2$ versus $H_a : p_1 \neq p_2$ at level $\alpha = .05$ at $n_1 = 40, p_1 = .3, n_2 = 50, p_2 = .1, ..., 9$ and plot the power curve.

Chapter 9

Inference for Two-Way Tables

New Minitab commands discussed in this chapter

Stat ▶ Tables ▶ Chi-Square Test

Stat ▶ Tables ▶ Cross Tabulation

In this chapter, inference methods are discussed for comparing the distributions of a categorical variable for a number of populations and for looking for relationships among a number of categorical variables defined on a single population. The *chi-square test* is the basic inferential tool, and this is implemented in Minitab via the Stat ▶ Tables ▶ Cross Tabulation command, if the data is in the form of raw incidence data, or the Stat ▶ Tables ▶ Chi-Square Test command, if the data comes in the form of counts.

9.1 Tabulating and Plotting

The relationship between two categorical variables is typically assessed by cross-tabulating the variables in a table. For this, the Stat ▶ Tables ▶ Cross Tabulation command is available. We illustrate using an example where each categorical variable takes two values. Of course, each variable can take a number of values, and this need not be the same for each categorical variable.

Suppose that we have collected data on courses being taken by students and have recorded a 1 in C2 if the student is taking Statistics and a 0 if not. If the student is taking Calculus, a 1 is recorded in C3 and a 0 otherwise. Also, we have recorded the student number in C1. These data for 10 students follow.

Row	C1	C2	C3
1	12389	1	0
2	97658	1	0
3	53546	0	1
4	55542	0	1
5	11223	1	1
6	77788	0	0
7	44567	1	1
8	32156	1	0
9	33456	0	1
10	67945	0	1

We cross-tabulate the data in C2 and C3 using the Stat ▶ Tables ▶ Cross Tabulation command and the dialog box shown in Display 9.1.

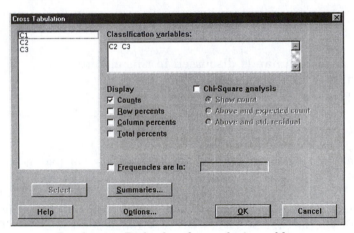

Display 9.1: Dialog box for producing tables.

This produces the output

```
Rows:  C2     Columns:  C3

              0         1       All

  0           1         4         5
  1           3         2         5
All           4         6        10


   Cell Contents --
                    Count
```

in the Session window that reveals there is 1 student taking neither Statistics nor Calculus, 4 students taking Calculus but not Statistics, 3 students taking Statistics but not Calculus, and 2 students taking both subjects. The row marginal totals are produced on the right, and the column marginal totals are produced below the table. We have chosen the cell entries in the table to be frequencies (counts) but we can see from Display 9.1 that there are other choices. For example, if we had checked the Total percents box instead, we obtain the output

```
Rows:  C2        Columns:  C3

             0        1         All

0         10.00    40.00      50.00
1         30.00    20.00      50.00
All       40.00    60.00     100.00

Cell Contents --
              % of Tbl
```

where each entry is the percentage that cell represents of the total number of observations used to form the table. Of course, we can ask for more than just one of these cell statistics to be produced in a table.

To examine the relationship between the two variables, we compare the conditional distributions given row, by checking the Row percents box, or the conditional distributions given column, by checking the Column percents box. For example, choosing to calculate row percents gives us the table

```
Rows:  C2        Columns:  C3

             0        1         All

0         20.00    80.00     100.00
1         60.00    40.00     100.00
All       40.00    60.00     100.00

Cell Contents --
              % of Row
```

that gives the row distributions as 20%, 80% for the first row and 60%, 40% for the second row. So it looks as if there is a strong relationship between the variable indicating whether or not a student takes Statistics and the variable indicating whether or not a student takes Calculus. For example, a student who does not take Statistics is more likely to take Calculus than a student who does take Statistics. Of course, this is not a real data set, and it is small at that. So, in reality, we could expect a somewhat different conclusion.

Some graphical techniques are also available for this problem. In Figure 9.1, we have plotted the conditional distributions given row in a bar chart using the command Graph ▶ Charts and filling in the dialog box so that Function is Count, the Y-variable is C3, the X-variable is C2, Display is Bar, and under Options we selected Cluster with variable C3 and Total Y to 100% within each X-category. The bars for C3 are ordered according to the increasing value of

C2. If you would rather there be a single bar for each category of the X-variable and this bar be subdivided according to the conditional distribution of the Y-variable, then, rather than Cluster with variable Y, use Stack with variable Y. These plots are an evocative way to display the relationship between the variables.

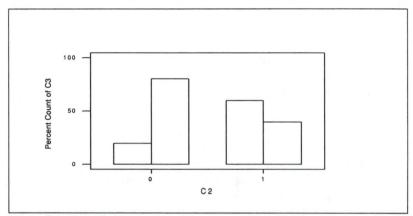

Figure 9.1: Conditional distributions of columns given row for the example of II.9.1.

The corresponding session command is **table** and there are the subcommands **totpercents, rowpercents,** and **colpercents** to specify whether or not we want total percents, row percents, and column percents to be printed for each cell. For example,

```
MTB > table c2 c3
```

produces the table of counts above. If you do not want the marginal statistics to be printed, use the **noall** subcommand. Any cases with missing values are not included in the cross-tabulation. If you want them to be included, use the **missing** subcommand and a row or column is printed, whichever is relevant, for missing values. For example, the subcommand

```
SUBC> missing c2 c3;
```

ensures that any cases with missing values in C2 or C3 are also tabulated.

9.2   The Chi-square Test

If you have a single variable, you can use the $\underline{S}$tat ▶ $\underline{T}$ables ▶ $\underline{C}$ross Tabulation command to form the table of counts if your data is not in this form. To carry out a chi-square goodness-of-fit test, however, you will have to use Minitab commands to directly compute the *chi-square statistic*

$$X^2 = \sum_{\text{cell}} \frac{(\text{observed count in cell} - \text{expected count in cell})^2}{\text{expected count in cell}}$$

and the P-value given by the probability

$$P(Y > X^2)$$

where Y follows a *Chisquare* (k) distribution based on an appropriate degrees of freedom k as determined by the table and the model being fitted. This is an approximate distribution result. Recall that the *Chisquare* (k) distribution was discussed in II.6.5.

When we have more than one variable and we are interested in whether or not a relationship exists, there are Minitab commands to carry out the chi-square analysis. Recall that there is no relationship between the variables — i.e., the variables are *independent* — if and only if the conditional distributions of one variable given the other are all the same. So in a two-way table we can assess whether or not there is a relationship by comparing the conditional distributions of the columns given the rows. Of course, there will be differences in these conditional distributions simply due to sampling error. Whether or not these differences are significant is assessed by conducting a chi-square test. When the table has r rows and c columns and we are testing for independence, then $k = (r-1)(c-1)$. Note that for a cell the square of a *cell's standardized residual* is that cell's contribution to the chi-square statistic, namely

$$\frac{(\text{observed count in cell } - \text{ expected count in cell})^2}{\text{expected count in cell}}$$

For example, suppose for 60 cases we have a categorical variable in C1 taking the values 0 and 1 and a categorical variable in C2 taking the values 0, 1 and 2. Suppose further that the $\underline{S}$tat ▶ $\underline{T}$ables ▶ $\underline{C}$ross Tabulation command with the dialog box as in Display 9.2

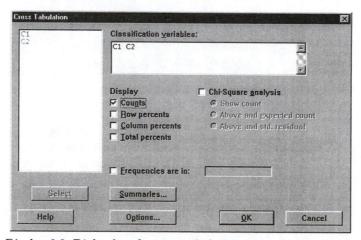

Display 9.2: Dialog box for cross-tabulating categorical variables.

produced the table

```
Rows:  C1     Columns:  C2

               0      1      2      All

0             10     13     11      34
1              9     10      7      26
All           19     23     18      60

Cell Contents --
                    Count
```

in the Session window. This records the counts in the 6 cells of a table with C1 indicating row and C2 indicating column. The variable C1 could be indicating a population with C2 a categorical variable defined on each population (or conversely), or both variables could be defined on a single population.

When using the Stat ▶ Tables ▶ Cross Tabulation command, a chi-square analysis can be carried out by checking the Chi-Square analysis box in the dialog box of Display 9.2. Actually, when we do this the appearance of this dialog box changes to that shown in Display 9.3.

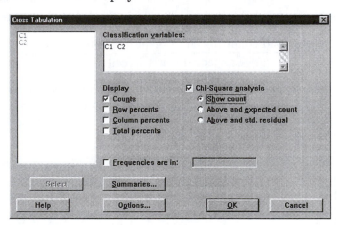

Display 9.3: Dialog box for carrying out a chi-square analysis from within the Stat ▶ Tables ▶ Cross Tabulation command.

We see that there are now several additional options concerning what is printed in the Session window. We have chosen to have only the cell count printed for each cell. In addition, the Chi-square statistic is printed and its associated P-value. Clicking on the OK button in this dialog box leads to the output

```
Rows:   C1     Columns:   C2
               0        1        2      All
0             10       13       11       34
1              9       10        7       26
All           19       23       18       60
Chi-Square = 0.271, DF = 2, P-Value = 0.873
Cell Contents --
                    Count
```

being printed in the Session window. The *P*-value for testing the null hypothesis that these two categorical variables are independent against the alternative that they are not independent is .873, and so we do not reject the null hypothesis.

It is possible to cross-tabulate more than two variables and to test simultaneously for mutual statistical independence among the variables using the Ṣtat ▶ Ṭables ▶ Ċross Tabulation command. Recall that it is also a good idea to plot the conditional distributions as well, and this can be carried out using Ġraph ▶ Ċharts and filling in the dialog box as described in II.9.1.

The general syntax of the corresponding session command **table** command is

table $E_1 \ldots E_m$;
chisquare V.

where E_1, ..., E_m are columns containing categorical variables and V is either omitted or takes the value 1, 2, or 3. The value 1 is the default and causes the count to be printed in each cell and can be omitted. The value 2 causes the count and the expected count, under the hypothesis of independence, to be printed in each cell. The value 3 causes the count, the expected count, and the standardized residual to be printed in each cell. For example, the command

```
MTB > table c1 c2;
SUBC> chisquare.
```

also produces the above output.

9.3   Analyzing Tables of Counts

If you have a two-way cross-tabulation for which the cell counts are already tabulated, you can use the Ṣtat ▶ Ṭables ▶ Chi-Square Ṭest command on this data to carry out the chi-square analysis. For example, suppose we put the following data in columns C1–C3 as

```
Row     C1      C2      C3
  1     51      22      43
  2     92      21      28
  3     68       9      22
```

corresponding to the counts arising from the cross-classification of a row and column variable. We then use the command Ṣtat ▶ Ṭables ▶ Chi-Square Ṭest on this data with the dialog box as shown in Display 9.4.

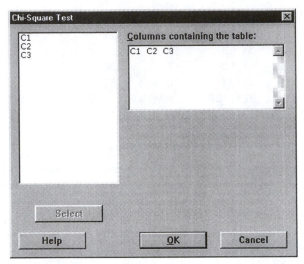

Display 9.4: Dialog box for chi-square test on a table of counts.

This produces the output

```
Expected counts are printed below observed counts
          C1      C2      C3      Total
    1     51      22      43       116
        68.75   16.94   30.30
    2     92      21      28       141
        83.57   20.60   36.83
    3     68       9      22        99
        58.68   14.46   25.86
 Total   211      52      93       356
Chi-Sq = 4.584 + 1.509 + 5.320 +
  0.850 + 0.008 + 2.119 +
  1.481 + 2.062 + 0.577 = 18.510
DF = 4, P-Value = 0.001
```

in the Session window. The chi-square statistic has the value 18.51 in this case and the *P*-value is .001, so we reject the null hypothesis that there is no relationship between the row and column variables.

The general syntax of the corresponding session command **chisquare** command is

chisquare $E_1 \ldots E_m$

and this computes the expected cell counts, the chi-square statistic, and the associated *P*-value for the table in columns E_1, ..., E_m. Note that there is a limitation on the number of columns; namely we must have $m \leq 7$. For example, the command

```
MTB > chisquare c1-c3
```

produces the above chi-square analysis.

9.4 Exercises

When the data for an exercise come from an exercise in IPS, the IPS exercise number is given in parentheses ( ). All computations in these exercises are to be carried out using Minitab, and the exercises are designed to ensure that you have a reasonable understanding of the Minitab material in this chapter. Generally, you should be using Minitab to do all the computations and plotting required for the problems in IPS.

1. Suppose that the observations in the following table are made on two categorical variables where variable 1 takes 2 values and variable 2 takes 3 values. Using the Stat ▶ Tables ▶ Cross Tabulation command, cross-tabulate this data in a table of frequencies and in a table of relative frequencies. Calculate the conditional distributions of variable 1, given variable 2. Plot the conditional distributions. Is there any indication of a relationship existing between the variables? How many conditional distributions of variable 2, given variable 1, are there?

Obs	1	2	3	4	5	6	7	8	9	10
Var 1	0	0	0	1	1	0	1	0	0	1
Var 2	2	1	0	0	2	1	2	0	1	1

2. (9.1) Use Minitab commands to calculate the marginal distributions and the conditional distributions, given Age group. Note that you cannot use Stat ▶ Tables ▶ Cross Tabulation for this. Plot the conditional distributions.

3. Use Minitab to directly compute the expected frequencies, standardized residuals, chi-square statistic, and P-value for the hypothesis of independence in the table of Example 9.17 in IPS.

4. (9.14) Carry out a chi-square analysis to determine whether or not the variables in this problem are related. Plot bar charts of the conditional distributions. Make sure you use the same scale on each plot so that they are comparable.

5. (9.15) Calculate and compare the conditional distributions of marital status given age. Plot these conditional distributions in bar charts.

6. Suppose we have a discrete distribution on the integers $1, \ldots, k$ with probabilities $p_1, \ldots, p_k$. Further, suppose we take a sample of n from this distribution and record the counts $f_1, \ldots, f_k$, where f_i records the number of times we observed i. It can be shown that

$$P(f_1 = n_1, \ldots, f_k = n_k) = \frac{n!}{n_1! \cdots n_k!} p_1^{n_1} \cdots p_k^{n_k}$$

when the n_i are nonnegative integers that sum to n. This is called the *Multinomial*$(n, p_1, \ldots, p_k)$ distribution, and it is a generalization of the

Binomial(n, p) distribution. It is the relevant distribution for describing the counts in cross-tabulations. For $k = 4, p_1 = p_2 = p_3 = p_4 = .25, n = 3$, calculate these probabilities and verify that it is a probability distribution. Note that the gamma function is available with the Çalc ▶ Calculator command (see Appendix B.1), and this can be used to evaluate factorials such as $n!$ and also $0! = 1$.

7. Calculate $P(f_1 = 3, f_2 = 5, f_3 = 2)$ for the *Multinomial*$(10, .2, .5, .3)$ distribution.

8. Generate (f_1, f_2, f_3) from the *Multinomial*$(1000, .2, .4, .4)$ distribution. Hint: Generate a sample of 1000 from the discrete distribution on 1, 2, 3 with probabilities .2, .4 , .4, respectively.

Chapter 10

Inference for Regression

New Minitab command discussed in this chapter

Stat ▶ Regression ▶ Residual Plots

This chapter deals with inference for the simple regression model. A regression analysis can be carried out using the command Stat ▶ Regression ▶ Regression. The regression as well as a scatterplot with the least-squares line overlaid can be obtained via Stat ▶ Regression ▶ Fitted Line Plot. Some aspects of these commands were discussed in II.2.3. Residual plots can be obtained using Stat ▶ Regression ▶ Residual Plots, provided you have saved the residuals.

10.1   Simple Regression Analysis

The command Stat ▶ Regression ▶ Regression provides a fit of the model $y = \alpha + \beta x + \epsilon$. Here, y is the *response variable*, x is the *explanatory* or *predictor variable*, ϵ is the *error variable* with an $N(0, \sigma)$ distribution, and α, β, and, σ are fixed unknown constants. These assumptions imply that, given x, the distribution of y is $N(\alpha + \beta x, \sigma)$. So the mean of y given x is $\alpha + \beta x$, and this gives the relationship between y and x, i.e., as x changes at most the mean of the conditional distribution of y given x changes according to the linear function $\alpha + \beta x$.

The primary aim of a regression analysis is to make inferences about the unknown intercept α and the unknown slope β and to make predictive inferences about future values of y at possibly new values of x. All inferences are dependent on this model being correct. If we go ahead and report inferences when the model is incorrect, we run the risk of these inferences being invalid. So we must always check that the model makes sense in light of the data obtained. This is referred to as *model checking*.

We let $(x_1, y_1), \ldots, (x_n, y_n)$ denote the data on which we will base all our

145

inferences. The basic inference method for this model is to use least-squares to estimate α and β and we denote these estimates by a and b, respectively, i.e., a and b are the values of α and β that minimize

$$S^2 = \sum_{i=1}^{n} (y_i - \alpha - \beta x_i)^2.$$

We predict the value of a future y, when the explanatory variable takes the value x, by $\hat{y} = a + bx$. The *ith fitted value* $\hat{y}_i$ is the estimate of the mean of y at x_i; — i.e., $\hat{y}_i = a + bx_i$. The *ith residual* is given by $r_i = y_i - \hat{y}_i$, i.e., it is the error incurred when predicting the value of y at x_i by $\hat{y}_i$. We estimate the standard deviation σ by

$$s = \sqrt{\frac{1}{n-2} \sum_{i=1}^{n} (y_i - \hat{y}_i)^2} = \sqrt{\frac{1}{n-2} \sum_{i=1}^{n} r_i^2}$$

which equals the square root of the MSE (mean-squared error) for the regression model.

Of course, the estimates a, b, and $\hat{y}$ are not equal to the quantities that they are estimating. It as an important aspect of a statistical analysis to say something about how accurate these estimates are, and for this we use the standard error of the estimate. The standard error of a is given by

$$s\sqrt{\frac{1}{n} + \frac{\bar{x}^2}{\sum_{i=1}^{n} (x_i - \bar{x})^2}}.$$

The standard error of b is given by

$$s\sqrt{\frac{1}{\sum_{i=1}^{n} (x_i - \bar{x})^2}}.$$

The standard error of the estimate $a + bx$ of the mean $\alpha + \beta x$ is given by

$$s\sqrt{\frac{1}{n} + \frac{(x - \bar{x})^2}{\sum_{i=1}^{n} (x_i - \bar{x})^2}}.$$

To predict y at x, we must take into account the additional variation caused by the error ϵ, and so the standard error of $a + bx$ as a predictor of y at x is given by

$$s\sqrt{1 + \frac{1}{n} + \frac{(x - \bar{x})^2}{\sum_{i=1}^{n} (x_i - \bar{x})^2}}.$$

Finally, the residual r_i, as an estimate of the error incurred at x_i, has standard error

$$s\sqrt{1 - \frac{1}{n} - \frac{(x_i - \bar{x})^2}{\sum_{i=1}^{n}(x_i - \bar{x})^2}}.$$

The *ith standardized residual* is then given by r_i divided by this quantity.

We now illustrate regression analysis using Minitab. Suppose we have the following data points

$$(x_1, y_1) = (1966, 73.1)$$
$$(x_2, y_2) = (1976, 88.0)$$
$$(x_3, y_3) = (1986, 119.4)$$
$$(x_4, y_4) = (1996, 127.1)$$

where x is year and y is yield in bushels per acre and that we give x the name **year** and y the name **yield**. The Stat ▶ Regression ▶ Fitted Line Plot command with **yield** as the response and **year** as predictor produces the plot of Display 10.1.

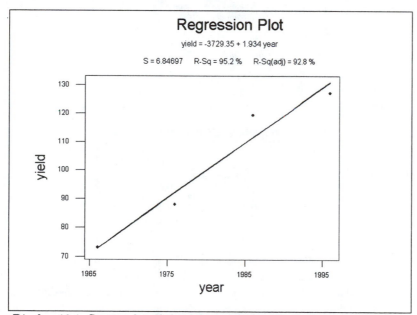

Display 10.1: Scatterplot of the data together with the least-squares line.

This command also produces some of the output below in the Session window. Because we wanted more features of a regression analysis than this command provides, we resorted to the Stat ▶ Regression ▶ Regression command together with the dialog boxes as in Displays 10.2, 10.3, 10.4, and 10.5.

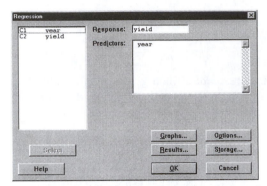

Display 10.2: Dialog box for simple regression analysis.

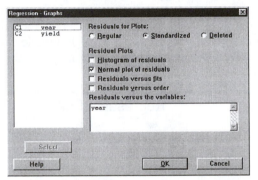

Display 10.3: Dialog box for selecting graphs to be plotted in regression analysis.

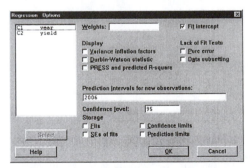

Display 10.4: Dialog box for selecting predictive inferences in a regression analysis.

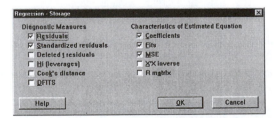

Display 10.5: Dialog box for selecting quantities to be stored in a regression analysis.

These produce the output

```
Regression Analysis
The regression equation is
yield = - 3729 + 1.93 year

Predictor      Coef     StDev       T         P
Constant    -3729.4     606.6    -6.15     0.025
year         1.9340    0.3062     6.32     0.024

S = 6.847 R-Sq = 95.2% R-Sq(adj) = 92.8%

Analysis of Variance
Source          DF       SS       MS        F        P
Regression       1    1870.2   1870.2    39.89    0.024
Residual Error   2      93.8     46.9
Total            3    1963.9

Predicted Values
    Fit   StDev Fit      95.0% CI           95.0% PI
  150.25       8.39  (114.17, 186.33)  (103.67, 196.83) X
X denotes a row with X values away from the center
```

in the Session window.

The dialog box of Display 10.2 establishes that yield is the response and year is the explanatory variable. The output from this gives the least-squares line as $y = -3729 + 1.93x$. Further, the standard error of $b_0 = -3729.4$ is 606.6, the standard error of $b_1 = 1.934$ is 0.3062, the t-statistic for testing $H_0 : \beta_0 = 0$ versus $H_a : \beta_0 \neq 0$ is -6.15 with P-value 0.025, and the t-statistic for testing $H_0 : \beta_1 = 0$ versus $H_a : \beta_1 \neq 0$ is 6.32 with P-value 0.024. The estimate of σ is $s = 6.847$ and the squared correlation — coefficient of determination — is $R^2 = .952$, indicating that 95% of the observed variation in y is explained by the changes in x. The Analysis of Variance table indicates that the F-statistic for testing $H_0 : \beta_1 = 0$ versus $H_a : \beta_1 \neq 0$ is 39.89 with P-value 0.024 and the MSE is 46.9. So we definitely reject the null hypothesis that there is no relationship between the response and the predictor.

Before clicking on the OK button of the dialog box of Display 10.2, however, we clicked on the Graphs button to bring up the dialog box of Display 10.3, the Options button to bring up the dialog box of Display 10.4, and the Storage button to bring up the dialog box of Display 10.5. In the Graphs dialog box, we specified that we wanted the standardized residuals plotted in a normal probability plot and plotted against the variable year. These plots appear in Displays 10.6 and 10.7, respectively, and don't indicate that any model assumptions are being violated.

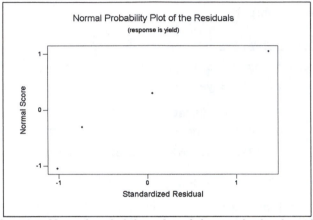

Display 10.6: Normal probability plot of the standardized residuals.

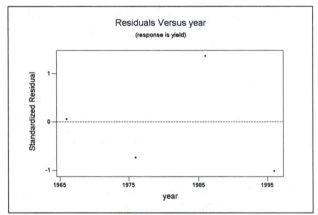

Display 10.7: A plot of the standardized residuals versus the explanatory variable.

In the Options dialog box, we specified that we wanted to estimate the mean value of y at $x = 2006$ and report and store this value together with a 95% confidence interval for this quantity and a 95% prediction interval for this quantity. The output above gives the estimated mean value at $x = 2006$ as 150.25 with standard error 8.39, and the 95% confidence and prediction intervals for this quantity are (114.17, 186.33) and (103.67, 196.83), respectively. The estimate is stored in the worksheet in a variable called `pfit1`, and the endpoints of the confidence and prediction intervals are stored in the worksheet with the names `clim1`, `clim2`, `plim1`, `plim2`, respectively. In the Storage dialog box, we specified that we wanted to store the values of a and b, the MSE, the fitted values, the residuals and the standardized residuals. The MSE is stored in the worksheet as a constant called `mse1`, the residuals are stored in a variable called `resi1`, the standardized residuals are stored in a variable called `sres1`, the values of a and b are stored consecutively in a variable named `coefs1`, and the fitted values are stored in a variable called `fits1`.

All of the stored quantities are available for further use. Suppose we want a 95% confidence interval for b. The commands

```
MTB > invcdf .975;
SUBC> student 2.
Student's t distribution with 2 DF
 P( X <= x)        x
    0.9750      4.3027
MTB > let k2=4.3027*.3062
MTB > let k3=coef1(2)-k2
MTB > let k4=coef1(2)+k2
MTB > print k3 k4
K3 0.616513
K4 3.25149
```

gives this interval as $(0.617, 3.251)$.

The general syntax of the corresponding session command **regress** command for fitting a line is

regress $E_1 1$ E_2

where E_1 contains the values of the response variable y and E_2 contains the values of the explanatory variable x. There are a number of subcommands that can be used with **regress**, and these are listed and explained below.

coefficients E_1 — stores the estimates of the coefficients in column E_1.

constant (noconstant) — ensures that β_0 is included in the regression equation, while **noconstant** fits the equation without β_0.

fits E_1 — stores the *fitted values* $\hat{y}$ in E_1.

ghistogram — causes a histogram of the residuals specified in **rtype** to be plotted.

gfits — causes a plot of the residuals specified in **rtype** versus the fitted values to be plotted.

gnormal — causes a normal quantile plot of the residuals specified in **rtype** to be plotted.

gorder — causes a plot of the residuals specified in **rtype** versus order to be plotted.

gvariable E_1 — causes a plot of the residuals specified in **rtype** versus the explanatory variable in column E_1 to be plotted.

mse E_1 — stores the mean squared error in constant E_1.

predict $E_1 \ldots E_k$ — (k is the number of explanatory variables where $k = 1$ with simple linear regression) computes and prints the predicted values at E_1, ..., E_k, where these are columns of the same length or constants with E_i corresponding to the *ith* explanatory variable. Also, this prints the estimated standard deviations of these values, confidence intervals for these values, and prediction intervals. The subcommand **predict** in turn has a number of subcommands.

confidence V — V specifies the level for the confidence intervals.

pfits E_1 — stores the predicted values in E_1.

psdfits E_1 — stores the estimated standard deviations of the predicted values in E_1.

climits E_1 E_2 — stores the lower and upper confidence limits for the predicted values in E_1 and E_2, respectively.

plimits E_1 E_2 — stores the lower and upper prediction limits for the predicted values in E_1 and E_2, respectively.

residuals E_1 — stores the regular residuals in E_1.

rtype V — indicates what type of residuals are to be used in the plotting subcommands, where V = 1 is the default and specifies regular residuals, V = 2 specifies standardized residuals, and V = 3 specifies Studentized deleted residuals.

sresiduals E_1 —stores the standardized residuals — the residuals divided by their estimated standard deviations — in E_1.

For example, the session commands

```
MTB > regress 'yield' 1 'year';
SUBC> coefficients c3;
SUBC> mse k1;
SUBC> fits c4;
SUBC> residuals c5;
SUBC> sresiduals c6;
SUBC> rtype 2;
SUBC> gnormal;
SUBC> gvariable 'year';
SUBC> predict 2006;
SUBC> pfits c7;
SUBC> climits c8 c9;
SUBC> plimits c10 c11.
```

produce the same results as the menu commands with the dialog boxes as in Displays 10.2, 10.3, 10.4, and 10.5 for the example of this section.

10.2   Exercises

When the data for an exercise come from an exercise in IPS, the IPS exercise number is given in parentheses ( ). All computations in these exercises are to be carried out using Minitab, and the exercises are designed to ensure that you have a reasonable understanding of the Minitab material in this chapter. Generally, you should be using Minitab to do all the computations and plotting required for the problems in IPS.

1. In C1, place the x values -3.0, -2.5, -2.0, ..., 2.5, 3.0. In C2, store a sample of 13 from the error ϵ, where ϵ is distributed $N(0, 2)$. In C3, store the values $y = \beta_0 + \beta_1 x + \epsilon = 1 + 3x + \epsilon$. Calculate the least-squares estimates of β_0 and β_1 and the estimate of σ^2. Repeat this example but take 5 observations at each value of x. Compare the estimates from the two situations and their estimated standard deviations.

2. In C1, place the x values -3.0, -2.5, -2.0, ..., 2.5, 3.0. In C2, store a sample of 13 from the error ϵ, where ϵ is distributed $N(0, 2)$. In C3, store the values $y = \beta_0 + \beta_1 x + \epsilon = 1 + 3x + \epsilon$. Plot the least-squares line. Repeat your computations twice after changing the first y observation to 20 and then to 50, and make sure the scales on all the plots are the same. What effect do you notice?

3. In C1, place the x values -3.0, -2.5, -2.0, ..., 2.5, 3.0. In C2, store a sample of 13 from the error ϵ, where ϵ is distributed $N(0, 2)$. In C3, store the values $y = \beta_0 + \beta_1 x + \epsilon = 1 + 3x + \epsilon$. Plot the standardized residuals in a normal quantile plot against the fitted values and against the explanatory variable. Repeat this, but in C3 place the values of $y = 1 + 3x - 5x^2 + \epsilon$. Compare the residual plots.

4. In C1, place the x values -3.0, -2.5, -2.0, ..., 2.5, 3.0. In C2, store a sample of 13 from the error ϵ, where ϵ is distributed $N(0, 2)$. In C3, store the values $y = \beta_0 + \beta_1 x + \epsilon = 1 + 3x + \epsilon$. Plot the standardized residuals in a normal quantile plot against the fitted values and against the explanatory variable. Repeat this, but in C2 place the values of a sample of 13 from the *Student*(1) distribution. Compare the residual plots.

5. In C1, place the x values -3.0, -2.5, -2.0, ..., 2.5, 3.0. In C2, store a sample of 13 from the error ϵ, where ϵ is distributed $N(0, 2)$. In C3, store the values $y = \beta_0 + \beta_1 x + \epsilon = 1 + 3x + \epsilon$. Calculate the predicted values and the lengths of .95 confidence and prediction intervals for this quantity at $x = .1, 1.1, 2.1, 3.5, 5, 10$, and 20. Explain the effect that you observe.

6. In C1, place the x values -3.0, -2.5, -2.0, ..., 2.5, 3.0. In C2, store a sample of 13 from the error ϵ, where ϵ is distributed $N(0, 2)$. In C3, store the values $y = \beta_0 + \beta_1 x + \epsilon = 1 + 3x + \epsilon$. Calculate the least-squares estimates and their estimated standard deviations. Repeat this, but for C1 the x values are to be 12 values of -3 and one value of 3. Compare your results and explain them.

Chapter 11

Multiple Regression

In this chapter we discuss *multiple regression;* i.e., we have a single numeric response variable y and $k > 1$ explanatory variables $x_1, \ldots, x_k$. There are no real changes in the behavior of the Stat ▶ Regression ▶ Regression command, and the descriptions we gave in Chapter 10 apply to this chapter as well. We present an example of a multiple regression analysis using Minitab. In Chapter 15, we show how a *logistic regression* — a regression where the response variable y is binary-valued — can be carried out using Minitab.

A multiple regression analysis can be carried out using Stat ▶ Regression ▶ Regression and filling in the dialog box appropriately. Residual plots can be obtained using Stat ▶ Regression ▶ Residual Plots provided you have saved the residuals. Also available in Minitab are *stepwise regression* using Stat ▶ Regression ▶ Regression ▶ Stepwise and *best subsets regression* using Stat ▶ Regression ▶ Regression ▶ Best Subsets.

11.1 Example

We consider a generated multiple regression example to illustrate the use of the Stat ▶ Regression ▶ Regression command in this context. Suppose that $k = 2$ and $y = \beta_0 + \beta_1 x_1 + \beta_2 x_2 + \epsilon = 1 + 2x_1 + 3x_2 + \epsilon$, where ϵ is distributed $N(0, \sigma)$ with $\sigma = 1.5$. We generated a sample of 16 from the $N(0, 1.5)$ distribution and placed these values in C1. In C2 we stored the values of x_1 and in C3 stored the values of x_2. Suppose that these variables take every possible combination of $x_1 = -1, -.5, .5, 1$ and $x_2 = -2, -1, 1, 2$. In C4, we placed the values of the response variable y.

We then proceeded to analyze this data as if we didn't know the values of β_0, β_1, β_2, and σ. The Stat ▶ Regression ▶ Regression command as implemented in Display 11.1, together with Displays 11.2 and 11.3,

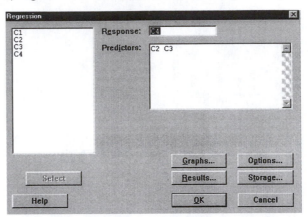

Display 11.1: Dialog box for the Stat ▶Regression ▶Regression command in the example.

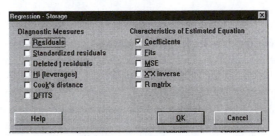

Display 11.2: Dialog box obtained by clicking on the Storage button in the dialog box depicted in Display 11.1. We have requested that the least-squares coefficients be stored.

Display 11.3: Dialog box obtained by clicking on the Options button in the dialog box depicted in Display 11.1. We have requested that a value be predicted at the settings $x_1 = 0, x_2 = 0$.

produces the output

```
The regression equation is
C4 = 1.00 + 2.38 C2 + 2.50 C3

Predictor      Coef     StDev       T        P
Constant     1.0014    0.3307    3.03    0.010
C2           2.3807    0.4183    5.69    0.000
C3           2.4964    0.2092   11.94    0.000

S = 1.323 R-Sq = 93.1% R-Sq(adj) = 92.0%

Analysis of Variance
Source              DF         SS       MS      F        P
Regression           2     305.95   152.98  87.42    0.000
Residual Error      13      22.75     1.75
Total               15     328.70

Source       DF     Seq SS
C2            1      56.68
C3            1     249.28

Unusual Observations
Obs    C2       C4      Fit   StDev Fit   Residual    St Resid
 6   -0.50   -5.574   -2.685     0.444     -2.889      -2.32R
R denotes an observation with a large standardized residual

Predicted Values
   Fit    StDev Fit       95.0% CI           95.0% PI
 1.001       0.331     (0.287, 1.716)    (-1.944, 3.947)

Values of Predictors for New Observations
New Obs          C2           C3
1          0.000000     0.000000
```

This specifies the least-squares equation as $y = 1.00 + 2.38x_1 + 2.50x_2$. For example, the estimate of β_1 is $b_1 = 2.3807$ with standard error 0.4183 and the t-statistic for testing $H_0 : \beta_1 = 0$ versus $H_a : \beta_1 \neq 0$ is 5.69 with P-value 0.000. The estimate of σ is $s = 1.323$ and $R^2 = .931$. The Analysis of Variance table indicates that the F-statistic for testing $H_0 : \beta_1 = \beta_2 = 0$ versus $H_a : \beta_1 \neq 0$ or $\beta_2 \neq 0$ takes the value 87.42 with P-value 0.000 so we would definitely reject the null hypothesis. Also, the MSE is given as 1.75.

The table after the Analysis of Variance table is called the *Sequential Analysis of Variance table* and is used when we want to test whether or not explanatory variables are in the model in a prescribed order. For example, the table that contains the rows labeled C2 and C3 allows for the testing of the sequence of hypotheses $H_0 : \beta_2 = 0$ versus $H_a : \beta_2 \neq 0$ and — if we reject this (and only if we do) — then testing the hypothesis $H_0 : \beta_1 = 0$ versus $H_a : \beta_1 \neq 0$. To test these hypotheses, we first compute $F = 249.28/s^2 = 249.28/1.75 = 142.45$ and then compute the P-value $P(F(1, 13) > 142.45) = 0.00$, and so we reject and go no further. If we had not rejected this null hypothesis, the second null hypothesis would be tested in exactly the same way. Obviously, the order in which we put variables into the model matters with these sequential tests. Sometimes, it is clear how to do this; e.g., in fitting a quadratic model $y = \beta_0 + \beta_1 x + \beta_2 x^2 + \epsilon$ we put $x_1 = x$ and $x_2 = x^2$ and test for the existence of the quadratic term first and, if no quadratic term is found, test for the existence of the linear term. Sometimes, the order for testing is not as clear and the sequential tests are not as appropriate.

The dialog box in Display 11.2 is obtained by clicking on the Storage button in the dialog box of Display 11.1. We stored the values of the least-squares estimates in C5, as the dialog box in Display 11.2 indicates, and so these are available for forming confidence intervals. Then, for example, the commands

```
MTB > invcdf .95;
SUBC> student 13.
Student's t distribution with 13 DF
 P(X <= x) x
 0.9500 1.7709
MTB > let k1=1.7709*.2092
MTB > let k2=c5(3)-k1
MTB > let k3=c5(3)+k1
MTB > print k2 k3
K2 2.12590
K3 2.86685
```

compute a 90% confidence interval for β_2 as $(2.126, 2.869)$, which we note does not cover the true value in this case.

The dialog box in Display 11.3 is obtained by clicking on the Options button in the dialog box of Display 11.1. The dialog box in Display 11.3 indicates that we requested that the program compute the predicted value at $x_1 = 0, x_2 = 0$ as well as the confidence and prediction intervals for this value. We obtained the predicted value as 1.001 with standard error .331 and as well the 95% confidence and prediction intervals given by $(0.287, 1.716)$ and $(-1.944, 3.947)$, respectively. Further, these limits were stored in the columns C6–C9.

The dialog box in Display 11.4 is obtained by clicking on the Graphs button in the dialog box of Display 11.1. Here we requested a normal quantile plot of the standardized residuals, which we show in Display 11.5, and also requested plots of the standardized residuals against each of the explanatory variables, which we don't show. All of these plots look reasonable although we note that the software has identified the sixth observation as having a large standardized residual even though we *know* that the model is correct. Of course, 16 is not many data points so we can expect inference to be somewhat unreliable in this case.

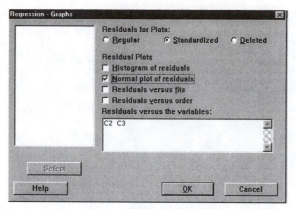

Display 11.4: Dialog box obtained by clicking on the Graphs button in the dialog box depicted in Display 11.1.

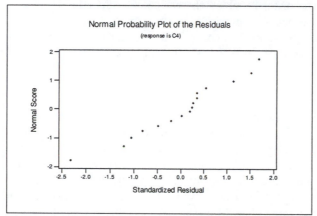

Display 11.5: Normal quantile plot of the standardized residuals for the example of Chapter 11.

The following session commands produce the above output for the example of this section.

```
MTB > regress c4 2 c2 c3;
SUBC> coefficients c5;
SUBC> rtype 2;
SUBC> gnormal;
SUBC> gvariable C2 C3;
SUBC> predict 0 0;
SUBC> climits c6 c7;
SUBC> plimits c8 c9.
```

We can also control the amount of output obtained from the Stat ▶ Regression ▶ Regression command. This is accomplished by clicking on the Results button of the dialog box shown in Display 11.1 bringing up Display 11.6.

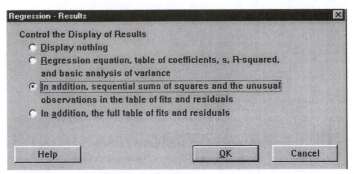

Display 11.6: Dialog box obtained by clicking on the Results button in the dialog box depicted in Display 11.1.

We have requested that, in addition to the fitted regression equation, least squares coefficients, s, R^2, and ANOVA table, the table of sequential sums of squares (for the order in which the variables appear in the model) and a table of unusual observations be printed.

The session command to control the amount output from the **regress** and other Minitab commands is **brief**. The general syntax of the **brief** command is

brief V

where V is a nonnegative integer that controls the amount of output. For any given command the output is dependent on the specific command although V = 0 suppresses all output, for all commands, beyond error messages and warnings. The default level of V is 2. When V = 3, the **regress** command produces the usual output and in addition prints x, y, $\hat{y}$, the standard deviation of $\hat{y}$, $y - \hat{y}$ and the standardized residual. When V = 1, the regress command gives the same output as when V = 2 but the sequential analysis of variance table is not printed. Don't forget that after you set the level of **brief,** this may affect the output of all commands you subsequently type and therefore it may need to be reset.

11.2   Exercises

When the data for an exercise come from an exercise in IPS, the IPS exercise number is given in parentheses ( ). All computations in these exercises are to be carried out using Minitab and the exercises are designed to ensure that you have a reasonable understanding of the Minitab material in this chapter. Generally, you should be using Minitab to do all the computations and plotting required for the problems in IPS.

1. In C1, place the x_1 values $-3.0, -2.5, -2.0, ..., 2.5, 3.0$. In C2, store a sample of 13 from the error ϵ, where ϵ is distributed $N(0, 2)$. In C3, store the values of $x_2 = x^2$. In C4 store the values of $y = \beta_0 + \beta_1 x_1 + \beta_2 x_2 + \epsilon = 1 + 3x + 5x^2 + \epsilon$. Calculate the least-squares estimates of β_0, β_1, and β_2

and the estimate of σ^2. Carry out the sequential F-tests testing first for the quadratic term and then, if necessary, testing for the linear term.

2. In C1, place the x values $-3.0, -2.5, -2.0, ..., 2.5, 3.0$. In C2, store a sample of 13 from the error ϵ, where ϵ is distributed $N(0, 2)$. Fit the model $y = 1 + 3\cos(x) + 5\sin(x) + \epsilon$. Calculate the least-squares estimates of β_0, β_1, and β_2 and the estimate of σ^2. Carry out the F-test for any effect due to x. Are the sequential F-tests meaningful here?

3. In C1, place the x_1 values $-3.0, -2.5, -2.0, ..., 2.5, 3.0$. In C2, store a sample of 13 from the error ϵ, where ϵ is distributed $N(0, 2)$. In C3, store the values of $x_2 = x^2$. In C4, store the values of $y = 1 + 3\cos(x) + 5\sin(x) + \epsilon$. Next fit the model $y = \beta_0 + \beta_1 x_1 + \beta_2 x_2 + \epsilon$ and plot the standardized residuals in a normal quantile plot and against each of the explanatory variables.

Chapter 12

One-Way Analysis of Variance

New Minitab commands discussed in this chapter

$\underline{S}$tat ► $\underline{A}$NOVA ► $\underline{O}$ne-way
$\underline{S}$tat ► $\underline{A}$NOVA ► One-way ($\underline{U}$nstacked)

This chapter deals with methods for making inferences about the relationship existing between a single numeric response variable and a single categorical explanatory variable. The basic inference methods are the one-way analysis of variance (ANOVA) and the comparison of means. There are two commands for carrying out a one-way analysis of variance, namely $\underline{S}$tat ► $\underline{A}$NOVA ► $\underline{O}$ne-way and $\underline{S}$tat ► $\underline{A}$NOVA ► One-way ($\underline{U}$nstacked). They differ in the way the data must be stored for the analysis.

We write the one-way ANOVA model as $x_{ij} = \mu_i + \epsilon_{ij}$, where $i = 1, \ldots, I$ indexes the levels of the categorical explanatory variable and $j = 1, \ldots, n_i$ indexes the individual observations at each level, μ_i is the mean response at the *ith* level, and the errors ϵ_{ij} are a sample from the $N(0, \sigma)$ distribution. Based on the observed x_{ij}, we want to make inferences about the unknown values of the parameters $\mu_1, \ldots, \mu_I, \sigma$

12.1 A Categorical Variable and a Quantitative Variable

Suppose that we have two variables, one variable is categorical and one is quantitative, and we want to examine the form of the relationship between these variables. Of course there may not even be a relationship between the variables. We treat the situation where the categorical variable is explanatory and

163

the quantitative variable is the response and examine some basic techniques for addressing this question.

To illustrate, we use the data in Exercise 2.18 of IPS. Here, we have four different colors of insect trap — lemon yellow, white, green, and blue — and the number of insects trapped on six different instances of each trap. The data are reproduced in the table below.

Board Color	Insects Trapped
Lemon Yellow	45 59 48 46 38 47
White	21 12 14 17 13 17
Green	37 32 15 25 39 41
Blue	16 11 20 21 14 7

We have read these data into a worksheet so that C1 contains the trap color, with 1 indicating lemon yellow, 2 indicating white, 3 indicating green, and 4 indicating blue, and in C2 we have put the numbers of insects trapped. We calculate the mean number of insects trapped for each trap using the Stat ▶ Tables ▶ Cross Tabulation command with the dialog boxes as in Displays 12.1 and 12.2. In the dialog box of Display 12.1 we have put C1 into the Classification variables box and clicked on the Summaries button to bring up the dialog box of Display 12.2. In this box, we have put C2 into the Associated variables box and selected Means to indicate that we want the mean of C2 to be computed for each value of C1.

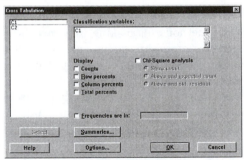

Display 12.1: First dialog box for tabulating a quantitative variable by a categorical variable.

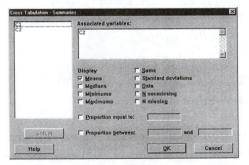

Display 12.2: Second dialog box for tabulating a quantitative variable by a categorical variable.

Clicking on the OK buttons produces the output

```
Rows:   C1

C2
Mean

1   47.833
2   15.667
3   31.500
4   14.833
All 27.292
```

in the Session window. The fact that the means change from one level of C1 to another seems to indicate that there is some relationship between the color of insect trap and the number of insects trapped. As indicated in Display 12.1, there are many other statistics, besides the mean, that we could have chosen to tabulate.

It is also a good idea to look at a scatterplot of the quantitative variable versus the categorical variable and to plot the means on this plot as well. We can do this with Graph ▶ Plot using the facility of that command to overlay scatterplots on the same plot. Here, we have put the values 1, 2, 3, 4 in C3 and the corresponding means in C4 and produced the plot shown in Display 12.3 with the means indicated by the symbol +.

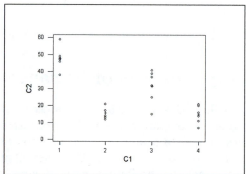

Display 12.3: Scatterplot of number of insects trapped versus trap color for Exercise 2.18 of IPS together with means for each color.

Another useful plot in this situation is to create side-by-side boxplots. This can be carried out using the Graph ▶ Boxplot command. The dialog box of Display 12.4 does this for the data of Exercise 2.18 of IPS as described above.

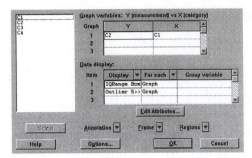

Display 12.4: Dialog box for creating side-by-side boxplots.

The session command **table** can also be used for creating the tables we have described in this section. For example, for the data of Exercise 2.18 of IPS as described above, the commands

```
MTB > table c1;
SUBC> means c2.
```

produce the mean number of insects trapped for each color of trap as given above. Besides the **means** subcommand, we have **medians, sums, minimums, maximums, n** (count of the nonmissing values), **nmiss** (count of the missing values), **stdev, stats** (equivalent to **n, means** and **stdev**), and **data** (lists the data for each cell). In addition, there is a subcommand **proportion** with the syntax

proportion $= V\ E_1$;

which gives the proportion of cases that have the value V in column E_1.

12.2   One-Way Analysis of Variance

The data in the table below (Example 12.6 of IPS) arose from a study of reading comprehension designed to compare three methods of instruction called basal, DRTA, and strategies. The data comprise scores on a test attained by children receiving each of the methods of instruction. There are 22 observations in each group. This study was conducted by Baumann and Jones of the Purdue School of Education.

Method	Scores
Basal	4 6 9 12 16 15 14 12 12 8 13 9 12 12 12 10 8 12 11 8 7 9
DRTA	7 7 12 10 16 15 9 8 13 12 7 6 8 9 9 8 9 13 10 8 8 10
Strat	11 7 4 7 7 6 11 14 13 9 12 13 4 13 6 12 6 11 14 8 5 8

We now carry out a one-way analysis of variance on this data to determine if there is any difference between the mean performances of students exposed to the three teaching methods. For this, we use the Stat ▶ ANOVA ▶ One-way command. For this example, there are $I = 3$ levels corresponding to the values Basal, DRTA, and Strat and $n_1 = n_2 = n_3 = 22$. Suppose that we have the values of the x_{ij} in C1 and the corresponding values of the categorical explanatory variable in C2, where Basal is indicated by 1, DRTA by 2, and Strat by 3. The Stat ▶ ANOVA ▶ One-way command together with the dialog boxes shown in Displays 12.6, 12.7, and 12.8 (described below) produce the output

```
Analysis of Variance for C1
Source    DF     SS      MS     F      P
C2         2    20.58   10.29  1.13   0.329
Error     63   572.45    9.09
Total     65   593.03

                        Individual 95% CIs For Mean
                        Based on Pooled StDev
Level  N   Mean  StDev  -----+---------+---------+---------+-
1      22 10.500  2.972       (---------*----------)
2      22  9.727  2.694     (----------*----------)
3      22  9.136  3.342 (-----------*----------)
                        -----+---------+---------+---------+-
Pooled StDev = 3.014         8.4       9.6      10.8      12.0
Fisher's pairwise comparisons
 Family error rate = 0.121
Individual error rate = 0.0500
Critical value = 1.998
Intervals for (column level mean) - (row level mean)
                  1               2
      2        -1.043
                2.589
      3        -0.452          -1.225
                3.180           2.407
```

in the Session window. The F-test in the ANOVA table with a P-value of .329 indicates that the null hypothesis in the hypothesis testing problem $H_0 : \mu_1 = \mu_2 = \mu_3$ versus $H_0 : \mu_1 \neq \mu_2$ or $\mu_1 \neq \mu_3$ would not be rejected. Also, the estimate of σ is given by $s = 3.014$ and 95% confidence intervals are plotted for the individual μ_i.

The dialog box of Display 12.6 carries out a one-way ANOVA for the data in C1, with the levels in C2, and puts the ordinary residuals in a variable called **resi1** and the fitted values in a variable called **fits1**. Note that because we assume a constant standard deviation and the number of observations is the same in each group, the ordinary residuals can be used in place of standardized residuals. Note also that the *ith* fitted value in this case is given by the mean of the group to which the observation belongs to.

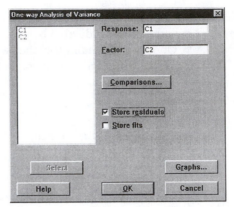

Display 12.6: Dialog box for one-way ANOVA.

The dialog box of Display 12.7 is obtained by clicking on the Comparisons button in the dialog box of Display 12.6. We use this dialog box to select a multiple comparison procedure.

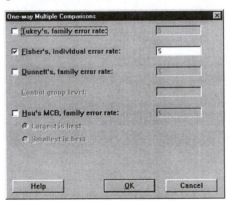

Display 12.7: Dialog box for selecting a multiple comparison procedure in a one-way ANOVA.

Here we have chosen to use the Fisher multiple comparison method with an individual error rate on the comparisons of 5%. This gives confidence intervals for the differences between the means using

$$\bar{x}_i - \bar{x}_j \pm s\sqrt{\frac{1}{n_i} + \frac{1}{n_j}}\, t^*$$

where s is the pooled standard deviation and t^* is the .975 percentile of the Student distribution with the error degrees of freedom. Note that with an individual 95% confidence interval, the probability of not covering the true difference (the *individual error rate*) is .05 but the probability of at least one of these three not covering the difference (the *family error rate*) is .121. If you want a more conservative family error rate, specify a lower individual error rate. For example, an individual error rate of .02 specifies a family error rate of .0516 in this example. We refer the reader to Help for details on the other available multiple comparison procedures. In the output above, we see that a 95% confidence interval for $\mu_1 - \mu_2$ is given by $(-1.043, 2.589)$, and because this includes 0, we conclude that there is no evidence against the null hypothesis $H_0 : \mu_1 = \mu_2$. We get the same result for the other two comparisons. Given that the F-test has already concluded that there is no evidence of any differences among the means, there is no reason for us to carry out these individual comparisons, and we do it only for illustration purposes here.

The dialog box of Display 12.8 is obtained by clicking on the Graphs button in the dialog box of Display 12.6. We have requested a plot of side-by-side boxplots of the data by level, which resulted in Display 12.9, a normal probability plot of the residuals that appears in Display 12.10 and a plot of the residuals against the index in C3 that appears in Display 12.11. The residual plots don't indicate any problems with the model assumptions.

Display 12.8: Dialog box for producing plots in a one-way ANOVA.

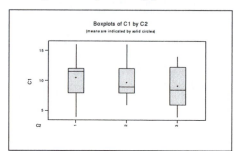

Display 12.8: Boxplots for the example of II.12.2.

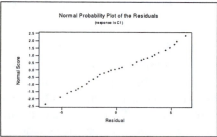

Display 12.9: Normal quantile plot for the example of II.12.2 after fitting a one-way ANOVA model.

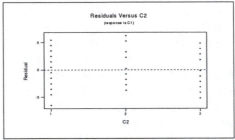

Display 12.10: Plot of residuals against level for the example of II.12.2 after fitting a
one-way ANOVA model.

A one-way ANOVA can also be carried out using $\underline{S}$tat ▶ $\underline{A}$NOVA ▶ One-way
($\underline{U}$nstacked) and filling in the dialog box appropriately. This command is much
more limited in its features than $\underline{S}$tat ▶ $\underline{A}$NOVA ▶ $\underline{O}$ne-way, however. So if
you have a worksheet with the samples for each level in columns, it would seem
better in general to use the $\underline{M}$anip ▶ $\underline{S}$tack command to place the data in one
column and then use $\underline{S}$tat ▶ $\underline{A}$NOVA ▶ $\underline{O}$ne-way.

Also available are analysis of means (ANOM) plots via $\underline{S}$tat ▶ $\underline{A}$NOVA ▶
A$\underline{n}$alysis of Means (see $\underline{H}$elp for details on these) and plots of the means with
error bars ($\pm$ one standard error of the observations at a level) via $\underline{S}$tat ▶
$\underline{A}$NOVA ▶ $\underline{I}$nterval Plots. Also, we can plot the means joined by lines using
$\underline{S}$tat ▶ $\underline{A}$NOVA ▶ $\underline{M}$ain Effects plots as in Display 12.11. The dotted line is
the grand mean. Power calculations can be carried out using $\underline{S}$tat ▶ $\underline{P}$ower and
Sample Size ▶ $\underline{O}$ne-way ANOVA and filling in the dialog box appropriately.

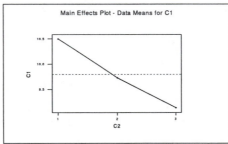

Display 12.11: Main effects plot for the example of II.12.2.

The corresponding session command is given by **onewayaov** and has the general
syntax

    onewayaov E_1 E_2 E_3 E_4

where E_1 is a variable containing the responses, E_2 is a variable containing
indices that indicate group membership, E_3 is a variable to hold the residuals,
and E_4 is a variable to hold the fitted values. Of course, E_3 and E_4 can be
dropped if they are not needed. There are various subcommands that can
be used. The **gboxplot** subcommand produces side-by-side boxplots. The
gnormal subcommand produces a normal probability plot of the residuals. The
gvariables E_1 subcommand results in a plot of the residuals against the variable

E_1. We could also obtain side-by-side dotplots of the data using the **gdotplot** subcommand, a histogram of the residuals using the **ghistogram** subcommand, a plot of the residuals against observation order using the **gorder** subcommand, and a plot of the residuals against the fitted values using the **gfits** subcommand. The **fisher** V_1 subcommand gives confidence intervals for the differences between the means, where V_1 is the individual error rate. Also available for multiple comparisons are the **tukey, dunnett,** and **mcb** subcommands. For example, the commands

```
MTB > onewayaov c1 c2 c3 c4;
SUBC> gboxplot;
SUBC> gnormal;
SUBC> gvariable c2;
SUBC> fisher.
```

result in the same output as we produced for the example of this section using the menu commands. Here the fits are stored in C4 and the residuals are stored in C3.

The **aovoneway** command can be used for a one-way ANOVA when the data for each level is in a separate column. For example, suppose that the three samples for the example of this section are in columns C3–C5. Then the command

```
MTB > aovoneway c3-c5
```

produces the same ANOVA table and confidence intervals for the means as **onewayaov.** Only a limited number of subcommands are available with this command, however.

12.3 Exercises

When the data for an exercise come from an exercise in IPS, the IPS exercise number is given in parentheses ( ). All computations in these exercises are to be carried out using Minitab, and the exercises are designed to ensure that you have a reasonable understanding of the Minitab material in this chapter. Generally, you should be using Minitab to do all the computations and plotting required for the problems in IPS.

1. Generate a sample of 10 from each of the $N(\mu_i, \sigma)$ distributions for $i = 1, \ldots, 5$, where $\mu_1 = 1, \mu_2 = 1, \mu_3 = 1, \mu_4 = 1, \mu_5 = 2$, and $\sigma = 3$. Carry out a one-way ANOVA and produce a normal probability plot of the residuals and the residuals against the explanatory variable. Compute .95 confidence intervals for the differences between the means. Compute an approximate set of .95 simultaneous confidence intervals for the differences between the means.

2. Generate a sample of 10 from each of the $N(\mu_i, \sigma_i)$ distributions for $i = 1, \ldots, 5$, where $\mu_1 = 1, \mu_2 = 1, \mu_3 = 1, \mu_4 = 1, \mu_5 = 2$, $\sigma_1 = \sigma_2 =$

$\sigma_3 = \sigma_4 = 3$, and $\sigma_5 = 8$. Carry out a one-way ANOVA and produce a normal probability plot of the residuals and the residuals against the explanatory variable. Compare the residual plots with those obtained in Exercise II.12.1.

3. The F-statistic in a one-way ANOVA, when the standard deviation σ is constant from one level to another, is distributed *noncentral* $F(k_1, k_2)$ with noncentrality λ, where $k_1 = I - 1$, $k_2 = n_1 + \cdots n_I - I$,

$$\lambda = \frac{\sum_{i=1}^{I} n_i \left(\mu_i - \bar{\mu} \right)^2}{\sigma^2}$$

and $\bar{\mu} = \sum_{i=1}^{I} n_i \mu_i / \sum_{i=1}^{I} n_i$. Using simulation, approximate the power of the test in Exercise II.12.1 with level .05 and the values of the parameters specified and compared your results with exact results obtained from Stat ▶ Power and Sample Size ▶ One-way ANOVA.

Chapter 13

Two-Way Analysis of Variance

New Minitab command discussed in this chapter

$\underline{S}$tat ▶ $\underline{A}$NOVA ▶ $\underline{T}$wo-way

This chapter deals with methods for making inferences about the relationship existing between a single numeric response variable and two categorical explanatory variables. The $\underline{S}$tat ▶ $\underline{A}$NOVA ▶ $\underline{T}$wo-way command is used to carry out a two-way ANOVA.

We write the two-way ANOVA model as $x_{ijk} = \mu_{ij} + \epsilon_{ijk}$, where $i = 1, \ldots, I$ and $j = 1, \ldots, J$ index the levels of the categorical explanatory variables and $k = 1, \ldots, n_{ij}$ indexes the individual observations at each treatment (combination of levels), μ_{ij} is the mean response at the *ith* level and the *jth* level of the first and second explanatory variable, respectively, and the errors ϵ_{ijk} are a sample from the $N(0, \sigma)$ distribution. Based on the observed x_{ijk}, we want to make inferences about the unknown values of the parameters $\mu_{11}, \ldots, \mu_{IJ}, \sigma$.

13.1 The Two-Way ANOVA Command

We consider a generated example, where $I = J = 2, \mu_{11} = \mu_{21} = \mu_{12} = \mu_{22} = 1, \sigma = 2$, and $n_{11} = n_{21} = n_{12} = n_{22} = 5$. The ϵ_{ijk} are generated as a sample from the $N(0, 2)$ distribution, and we put $x_{ijk} = \mu_{ij} + \epsilon_{ijk}$ for $i = 1, \ldots, I$ and $j = 1, \ldots, J$ and $k = 1, \ldots, n_{ij}$. Note that the $\underline{S}$tat ▶ $\underline{A}$NOVA ▶ $\underline{T}$wo-way command requires balanced data; i.e. all the n_{ij} must be equal. We pretend that we don't know the values of the parameters and carry out a two-way analysis of variance. If the x_{ijk} are in C1, the values of i in C2 and the values of j in C3, the dialog box of Display 13.1

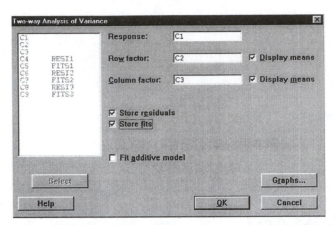

Display 13.1: Dialog box for producing a two-way analysis of variance.

produces the following output.

```
Analysis of Variance for C1
Source        DF      SS      MS     F      P
C2             1    0.39    0.39  0.07  0.790
C3             1    3.43    3.43  0.65  0.432
Interaction    1    4.01    4.01  0.76  0.396
Error         16   84.44    5.28
Total         19   92.26

               Individual 95% CI
C2   Mean  -+---------+---------+---------+---------+
1    1.49  (-------------------*------------------)
2    1.77      (------------------*------------------)
           -+---------+---------+---------+---------+
          0.00      0.80      1.60      2.40      3.20

               Individual 95% CI
C3   Mean  ----+---------+---------+---------+-------
1    2.05        (--------------*---------------)
2    1.22  (--------------*---------------)
           ----+---------+---------+---------+-------
              0.00      1.00      2.00      3.00
```

We see from this that the null hypothesis of no interaction is not rejected (P-value = .396) and neither is the null hypothesis of no effect due to the C2 factor (P-value = .790) nor the null hypothesis of no effect due to factor C3 (P-value = .432).

Note that by checking the Display means and Display means boxes in the dialog box of Display 13.1 we have caused 95% confidence intervals to be printed for the response means at each value of C2 and each value of C3, respectively. These cell means are relevant only when we decide that there is no interaction, as is the case here, and we note that all the intervals contain the true value 1 of these means.

We also checked the Store residuals and Store fits in the dialog boxes of Display 13.1. This results in the (ordinary) residuals being stored in C4 and the fitted values (cell means) being stored in C5. If these columns already had entries the next two available columns would be used instead.

If we want to fit the model without any interaction, supposing we know this to be true, we can check the Fit additive model box in the dialog box of Display 13.1. This is acceptable only in rare circumstances, however, as it is unlikely that we will know that this is true.

Various graphs are also available via the Graphs button in the dialog box of Display 13.1. Clicking on this results in the Dialog box shown in Display 13.2. Here we have asked for a normal probability plot of the (ordinary) residuals and a plot of the (ordinary) residuals versus the variables C2 and C3. Recall that with balance it is acceptable to use the ordinary residuals rather than the standardized residuals. We haven't reproduced the corresponding plots here but, as we might expect, they gave no grounds for suspecting the correctness of the model.

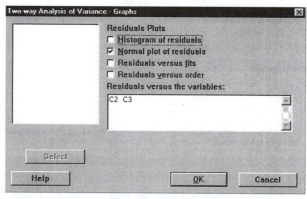

Display 13.2: Dialog box for producing various residual plots obtained via the Graphs button in the dialog box of Display 13.1.

If we conclude that there is an interaction then we must look at the individual IJ cell means to determine where the interaction occurs. A plot of these cell means is often useful in this regard. Also available are analysis of means (ANOM) plots via Stat ▶ ANOVA ▶ Analysis of Means. In addition, we can plot the marginal means joined by lines using Stat ▶ ANOVA ▶ Main Effects Plot and plot the cell means joined by lines using Stat ▶ ANOVA ▶ Interaction Plot using the dialog box of Display 13.3 with the output in Display 13.4.

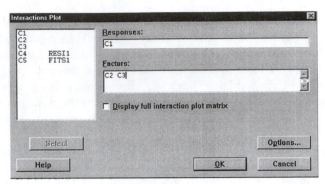

Display 13.3: Dialog box for obtaining the interaction plot of Display 13.4.

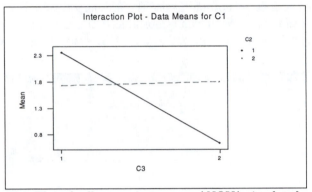

Display 13.4: Plot of cell means in two-way ANOVA simulated example.

Note that while the plot seems to indicate an interaction, this is not confirmed by the statistical test. Power calculations can be carried out using $\underline{S}$tat ▶ Power and Sample Size ▶ 2-Level $\underline{F}$actorial Design and filling in the dialog box appropriately. Commands are available in Minitab for analyzing unbalanced data and for situations where there are more than two factors where some factors are continuous and some categorical, and so on.

The corresponding session command for carrying out a two-way ANOVA is given by **twowayaov.** For example, the command

```
MTB > twowayaov c1 c2 c3 c4 c5;
SUBC> gnormal;
SUBC> gvariable c2 c3;
SUBC> means c2 c3.'
```

results in the same output as above. The **gnormal** subcommand results in a normal probability plot of the residuals being plotted while the **gvariables** subcommand results in a plot of the residuals against each of the factors C2 and C3. The **ghistogram, gfits,** and **gorder** subcommands are also available for a histogram of the residuals, the residuals against the fitted values, and the residuals against observation order, respectively. The **means** subcommand causes the estimates of marginal means for each level of C2 and C3 to be printed together with 95% confidence intervals. If we want to fit the model without any interaction, supposing we know this to be true, then the **additive** subcommand is available to do this.

13.2 Exercises

When the data for an exercise come from an exercise in IPS, the IPS exercise number is given in parentheses ( ). All computations in these exercises are to be carried out using Minitab and the exercises are designed to ensure that you have a reasonable understanding of the Minitab material in this chapter. Generally, you should be using Minitab to do all the computations and plotting required for the problems in IPS.

1. Suppose $I = J = 2, \mu_{11} = \mu_{21} = 1$ and $\mu_{12} = \mu_{22} = 2, \sigma = 2$, and $n_{11} = n_{21} = n_{12} = n_{22} = 10$. Generate the data for this situation, and carry out a two-way analysis. Plot the cell means (an interaction effect plot). Do your conclusions agree with what you know to be true?

2. Suppose $I = J = 2, \mu_{11} = \mu_{21} = 1$ and $\mu_{12} = 3, \mu_{22} = 2, \sigma = 2$ and $n_{11} = n_{21} = n_{12} = n_{22} = 10$. Generate the data for this situation, and carry out a two-way analysis. Plot the cell means (an interaction effect plot). Do your conclusions agree with what you know to be true?

3. Suppose $I = J = 2, \mu_{11} = \mu_{21} = 1$ and $\mu_{12} = \mu_{22} = 2, \sigma = 2$, and $n_{11} = n_{21} = n_{12} = n_{22} = 10$. Generate the data for this situation, and carry out a two-way analysis. Form 95% confidence intervals for the marginal means. Repeat your analysis using the additive model and compare the confidence intervals. Can you explain your results?

Chapter 14

Nonparametric Tests

New Minitab commands discussed in this chapter

Stat ▶ Nonparametrics ▶ Kruskal-Wallis
Stat ▶ Nonparametrics ▶ Mann-Whitney
Stat ▶ Nonparametrics ▶ 1-Sample Wilcoxon

This chapter deals with inference methods that do not depend upon the assumption of normality. These methods are sometimes called *nonparametric* or *distribution free* methods. Recall that we discussed a distribution-free method in II.7.4, where we presented the Stat ▶ Nonparametrics ▶ 1-Sample Sign command for the sign confidence interval and sign test for the median. Recall also the Manip ▶ Rank command in I.11.5, which can be used to compute the ranks of a data set.

14.1 The Wilcoxon Rank Sum Procedures

The Mann-Whitney test for a difference between the locations of two distributions is equivalent to the Wilcoxon rank sum test in the following sense. Suppose that we have two independent samples $y_{11}, \ldots, y_{1n_1}$ and $y_{21}, \ldots, y_{2n_2}$ from two distributions that differ at most in their locations as represented by their medians. The Mann-Whitney statistic U is the number of pairs (y_{1i}, y_{2j}) where $y_{1i} > y_{2j}$, while the Wilcoxon rank sum test statistic W is the sum of the ranks from the first sample when the ranks are computed for the two samples considered as one sample combined. It can be shown that $W = U + n_1(n_{1+1})/2$.

For Example 14.1 of IPS, we store the four values 166.7, 172.2, 165.0, and 176.9 of sample 1 in C1 and the four values 158.6, 176.4, 153.1, and 156.0 of sample 2 in C2. The Stat ▶ Nonparametrics ▶ Mann-Whitney command, implemented as in the dialog box of Display 14.1,

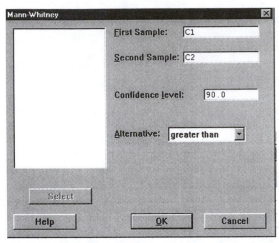

Display 14.1: Dialog box for implementing the Mann-Whitney command.

leads to the output,

```
Mann-Whitney Confidence Interval and Test
C1 N = 4 Median = 169.45
C2 N = 4 Median = 157.30
Point estimate for ETA1-ETA2 is 11.30
93.9 Percent CI for ETA1-ETA2 is (-9.70,20.90)
W = 23.0
Test of ETA1 = ETA2 vs ETA1 > ETA2 is significant at 0.0970
Cannot reject at alpha = 0.05
```

which indicates that the test of H_0 : the medians of the two distributions are identical versus H_a : the median of the first distribution is greater than the median of the second gives a P-value of .0970. Also, an estimate of 11.3 is produced for the difference in the medians, and we asked for a 90% confidence interval for this difference by placing **90** in the Confidence level box. Note that exact confidences cannot be attained due to the discrete distribution followed by the statistic U. The Mann-Whitney test requires the assumption that the two distributions we are sampling from have the same form.

The corresponding session command is given by **mann-whitney.** For example, the command

```
MTB > mann-whitney 90 c1 c2;
SUBC> alternative 1.
```

leads to the above output. Note that we have placed 90 on the command line to indicate that we want a 90% confidence interval. If this value is left out, a default 95% confidence interval is computed. Also available are the one-sided test of H_0 : the medians of the two distributions are identical versus H_a : the median of the first distribution is smaller than the median of the second, using the subcommand **alternative** -1 and the two-sided test is obtained if no **alternative** subcommand is employed.

14.2 The Wilcoxon Signed Rank Procedures

The Wilcoxon signed rank test and confidence interval are used for inferences about the median of a distribution. The Wilcoxon procedures are based on ranks, which is not the case for the sign procedures discussed in II.7.4. Consider the data of Example 14.8 in IPS where the differences between two scores has been recorded as $.37, -.23, .66, -.08, -.17$ in C1. The $\underline{\text{S}}$tat ▶ Nonparametrics ▶ 1-Sample $\underline{\text{W}}$ilcoxon command, implemented as in the dialog box in Display 14.2,

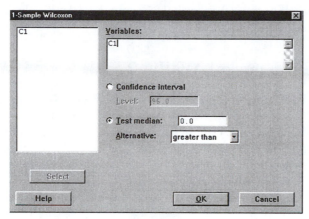

Display 14.2: Dialog box for implementing the Wilcoxon signed rank test.

leads to the output

```
Test of median = 0.000000 versus median > 0.000000
              N for    Wilcoxon              Estimated
       N     Test     Statistic      P       Median
C1     5      5          9.0      0.394       0.1000
```

which gives the P-value .394 for testing H_0 : the median of the difference is 0 versus H_a : the median of the difference is greater than 0. If instead we had filled in the $\underline{\text{C}}$onfidence interval button and placed 90 in the $\underline{\text{L}}$evel box of the dialog box in Display 14.2, we would have obtained the output

```
              Estimated    Achieved
       N       Median     Confidence    Confidence Interval
C1     5        0.100        89.4        (-0.200, 0.515)
```

which provides a 90% confidence interval for the median. Note that the Wilcoxon signed rank procedures for the median require an assumption that the response values (in this case the difference) come from a distribution symmetric about its median.

The corresponding session commands are given by **wtest** and **winterval** for tests and confidence intervals respectively. The general syntax of the **wtest** command is

wtest V E_1

where V is the hypothesized value of the median, with 0 being the default value, and E_1 is the column containing the data. For example, the command

```
MTB > wtest c1;
SUBC> alternative 1.
```

produces the above output for the test. The general syntax of the **winterval** command is

winterval V E_1

where V is the confidence level, with 0.95 being the default value, and E_1 is the column containing the data.

14.3   The Kruskal-Wallis Test

The Kruskal-Wallis test is the analog of the one-way ANOVA in the nonparametric setting. Suppose the data for Example 14.13 in IPS are in C1 and C2, where C1 contains the corn yield in bushels per acre and C2 is number of weeds per meter. The Stat ▶ Nonparametrics ▶ Kruskal-Wallis command, as implemented in Display 14.3,

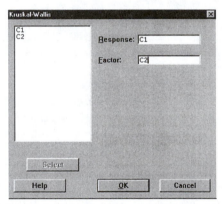

Display 14.3: Dialog box for implementing the Kruskal-Wallis test.

produces the output

```
Kruskal-Wallis Test on C1
C2       N   Median   Ave Rank      Z
0        4   169.4       13.1     2.24
1        4   163.6        8.4    -0.06
3        4   157.3        6.2    -1.09
9        4   162.6        6.2    -1.09
Overall 16 8.5

H = 5.56 DF = 3 P = 0.135
H = 5.57 DF = 3 P = 0.134 (adjusted for ties)
* NOTE * One or more small samples
```

which gives a P-value of .135 for testing H_0 : each sample comes from the same distribution versus H_a : at least two of the samples come from different distributions. Note that the validity of the Kruskal-Wallis test relies on the assumption that the distributions being sampled from all have the same form.

The corresponding session command is given by **kruskal-wallis.** For example, the command

```
MTB > kruskal-wallis c1 c2
```

also produces the above output. The general syntax of the **kruskal-wallis** command is

kruskal-wallis E_1 E_2

where E_1 contains the data and E_2 contains the levels of the factor.

14.4   Exercises

When the data for an exercise come from an exercise in IPS, the IPS exercise number is given in parentheses ( ). All computations in these exercises are to be carried out using Minitab and the exercises are designed to ensure that you have a reasonable understanding of the Minitab material in this chapter. Generally, you should be using Minitab to do all the computations and plotting required for the problems in IPS.

1. Generate a sample of $n = 10$ from the $N(0,1)$ distribution and compute the P-value for testing H_0 : the median is 0 versus H_a : the median is not 0, using the t-test and the Wilcoxon signed rank test. Compare the P-values. Repeat this with $n = 100$.

2. Generate a sample of $n = 10$ from the $N(0,1)$ distribution and compute 95% confidence intervals for the median, using the t-confidence interval and the Wilcoxon signed rank confidence intervals. Compare the lengths of the confidence intervals. Repeat this with $n = 100$.

3. Generate two samples of $n = 10$ from the $Student(1)$ distribution and add 1 to the second sample. Test H_0 : the medians of the two distributions are identical versus H_a : the medians are not equal using the two sample t-test and using the Mann-Whitney test. Compare the results.

4. Generate a sample of 10 from each of the $N(1,2), N(2,2)$, and $N(3,1)$ distributions. Test for a difference among the distributions using a one-way ANOVA and using the Kruskal-Wallis test. Compare the results.

5. Generate 10 scores for 10 brands from the $N(\mu_{ij}, \sigma)$ distributions for $i = 1, 2$ and $j = 1, 2$, where $\mu_{11} = \mu_{21} = 1$ and $\mu_{12} = \mu_{22} = 2$, and treat each test for no effect due to brand using a two-way ANOVA with the assumption of no interaction and also using the Friedman test. Compare the results.

Chapter 15

Logistic Regression

New Minitab commands discussed in this chapter

Stat ▶ Regression ▶ Binary Logistic Regression
Stat ▶ Regression ▶ Nominal Logistic Regression
Stat ▶ Regression ▶ Ordinal Logistic Regression

This chapter deals with the *logistic regression model*. This model arises when the response variable y is binary, i.e., takes only two values, and we have a number of explanatory variables $x_1, \ldots, x_k$.

15.1   The Logistic Regression Model

The regression techniques discussed in chapters 10 and 11 require that the response variable y be a continuous variable. In many contexts, however, the response is discrete and in fact binary; i.e., taking the values 0 and 1. Let p denote the probability of a 1. This probability is related to the values of the explanatory variables $x_1, \ldots, x_k$.

We cannot, however, write this as $p = \beta_0 + \beta_1 x_1 + \ldots + \beta_k x_k$ because the right-hand side is not constrained to lie in the interval [0,1], which it must if it is to represent a probability. One solution to this problem is to employ the *logit link function*, which is given by

$$\ln\left(\frac{p}{1-p}\right) = \beta_0 + \beta_1 x_1 + \cdots + \beta_k x_k$$

and this leads to the equations

$$\frac{p}{1-p} = \exp\{\beta_0 + \beta_1 x_1 + \cdots + \beta_k x_k\}$$

and

$$p = \frac{\exp\left\{\beta_0 + \beta_1 x_1 + \cdots + \beta_k x_k\right\}}{1 + \exp\left\{\beta_0 + \beta_1 x_1 + \cdots + \beta_k x_k\right\}}$$

for the *odds* $p/(1-p)$ and probability p, respectively. The right-hand side of the equation for p is now always between 0 and 1. Note that logistic regression is based on an ordinary regression relation between the logarithm of the odds in favor of the event occurring at a particular setting of the explanatory variables and the values of the explanatory variables $x_1, \ldots, x_k$. The quantity $\ln\left(p/(1-p)\right)$ is referred to as the *log odds*.

The procedure for estimating the coefficients $\beta_0, \beta_1, \ldots, \beta_k$ using this relation and carrying out tests of significance on these values is known as *logistic regression*. Typically, more sophisticated statistical methods than least squares are needed for fitting and inference in this context, and we rely on software such as Minitab to carry out the necessary computations.

In addition, other link functions are available in Minitab are often used. In particular, the *probit link function* is given by

$$\Phi^{-1}\left(p\right) = \beta_0 + \beta_1 x_1 + \cdots + \beta_k x_k$$

where Φ is the cumulative distribution function of the $N(0,1)$ distribution, and this leads to the relation

$$p = \Phi\left(\beta_0 + \beta_1 x_1 + \cdots + \beta_k x_k\right)$$

which is also always between 0 and 1. Choice of the link function can be made via a variety of goodness-of-fit tests available in Minitab, but we restrict our attention here to the logit link function.

15.2   Example

Suppose that we have the following 10 observations in columns C1–C3

Row	C1	C2	C3
1	0	−0.65917	0.43450
2	0	0.69408	0.48175
3	1	−0.28772	0.08279
4	1	0.76911	0.59153
5	1	1.44037	2.07466
6	0	0.52674	0.27745
7	1	0.38593	0.14894
8	1	−0.00027	0.00000
9	0	1.15681	1.33822
10	1	0.60793	0.36958

where the response y is in C1, x_1 is in C2, and x_2 is in C3 and note that $x_2 = x_1^2$. We want to fit the model

$$\ln\left(\frac{p}{1-p}\right) = \beta_0 + \beta_1 x_1 + \beta_2 x_2$$

and conduct statistical inference concerning the parameters of the model.

Fitting and inference is carried out in Minitab using $\underline{S}$tat ▶ $\underline{R}$egression ▶ Binary $\underline{L}$ogistic Regression and filling in the dialog box as in Display 15.1.

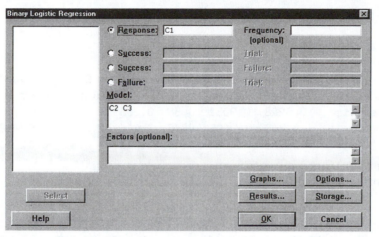

Display 15.1: Dialog box for implementing a binary logistic regression.

Here, the $\underline{R}$esponse box contains C1 and the $\underline{M}$odel box contains C2 and C3. Clicking on the $\underline{R}$esults button brings up the dialog box in Display 15.2.

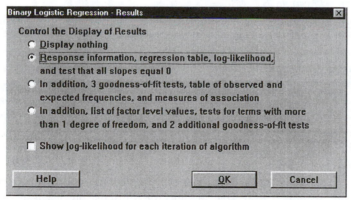

Display 15.2: The dialog box resulting from clicking on the $\underline{R}$esults button in the dialog box of Display 15.1.

We have filled in the radio button $\underline{R}$esponse information, regression table, etc., as this controls the amount of output. The default output is more extensive and we chose to limit this. The following output is obtained.

```
Link Function:  Logit
Response Information

Variable Value Count
C1        1        6  (Event)
          0        4
        Total     10
Logistic Regression Table
                                         Odds      95% CI
Predictor        Coef   StDev     Z     P   Ratio Lower Upper
Constant       0.5228  0.9031  0.58 0.563
C2              0.740   1.605  0.46 0.645   2.10  0.09 48.71
C3             -0.780   1.584 -0.49 0.623   0.46  0.02 10.23

Log-Likelihood = -6.598
Test that all slopes are zero:  G = 0.265, DF = 2,
                                         P-Value = 0.876
```

This gives estimates of the coefficients and their standard errors and the P-value for $H_0 : \beta_0 = 0$ versus $H_a : \beta_0 \neq 0$ as 0.563, the P-value for $H_0 : \beta_1 = 0$ versus $H_a : \beta_1 \neq 0$ as 0.643, and the P-value for $H_0 : \beta_2 = 0$ versus $H_a : \beta_2 \neq 0$ as 0.623. Further, the test of $H_0 : \beta_1 = \beta_2 = 0$ versus $H_a : \beta_1 \neq 0$ or $\beta_2 \neq 0$ has P-value .876. In this example, there is no evidence of any nonzero coefficients. Note that $p = .5$ when $\beta_0 = \beta_1 = \beta_2 = 0$

Also provided in the output is the estimate 2.10 for the odds ratio for x_1 (C2) and a 95% confidence interval $(.09, 48.71)$ for the true value. The odds ratio for x_1 is given by $\exp(\beta_1)$, which is the ratio of the odds at $x_1 + 1$ to the odds at x_1 when x_2 is held fixed or when $\beta_2 = 0$. Because there is evidence that $\beta_2 = 0$ (P-value $= .623$), the odds ratio has a direct interpretation here. Note, however, that if this wasn't the case the odds ratio would not have such an interpretation as it doesn't makes sense for x_2 to be held fixed when x_1 changes in this example as they are not independent variables. Similar comments apply to the estimate 0.46 for the odds ratio for x_2 (C3) and the 95% confidence interval $(.02, 10.23)$ for the true value of this quantity.

Many other aspects of fitting logistic regression models are available in Minitab and we refer the reader to Help for a discussion of these. Also available in Minitab are *ordinal logistic regression*, when the response takes more than two values and these are ordered, and *nominal logistic regression*, when the response takes more than two values and these are unordered. These can be accessed via Stat ▶ Regression ▶ Ordinal Logistic Regression and Stat ▶ Regression ▶ Nominal Logistic Regression, respectively.

15.3   Exercises

When the data for an exercise come from an exercise in IPS, the IPS exercise number is given in parentheses ( ). All computations in these exercises are to be carried out using Minitab and the exercises are designed to ensure that you have a reasonable understanding of the Minitab material in this chapter. Generally, you should be using Minitab to do all the computations and plotting required for the problems in IPS.

1. Generate a sample of 20 from the *Bernoulli*(.25) distribution. Pretending that we don't know p, compute a 95% confidence interval for this quantity. Using this confidence interval, form 95% confidence intervals for the odds and the log odds.

2. Let x take the values -1, $-.5$, 0, $.5$, and 1. Plot the log odds

$$\ln\left(\frac{p}{1-p}\right) = \beta_0 + \beta_1 x$$

against x when $\beta_0 = 1$ and $\beta_1 = 2$. Plot the odds and the probability p against x.

3. Let x take the values -1, $-.5$, 0, $.5$, and 1. At each of these values, generate a sample of 4 values from the *Bernoulli*(p_x) distribution where

$$p_x = \frac{\exp\{1 + 2x\}}{1 + \exp\{1 + 2x\}}$$

and let these values be the y response values. Carry out a logistic regression analysis of this data using the model.

$$\ln\left(\frac{p_x}{1-p_x}\right) = \beta_0 + \beta_1 x$$

Compute a 95% confidence interval for β_1 and determine if it contains the true value. Similarly, form a 95% confidence interval for the odds ratio when x increases by 1 unit and determine if it contains the true value.

4. Let x take the values -1, $-.5$, 0, $.5$, and 1. At each of these values, generate a sample of 4 values from the *Bernoulli*(p_x) distribution where

$$p_x = \frac{\exp\{1 + 2x\}}{1 + \exp\{1 + 2x\}}$$

and let these values be the y response values. Carry out a logistic regression analysis of this data using the model

$$\ln\left(\frac{p_x}{1-p_x}\right) = \beta_0 + \beta_1 x + \beta_2 x^2$$

Test the null hypothesis $H_0 : \beta_2 = 0$ versus $H_a : \beta_2 \neq 0$. Form a 95% confidence interval for the odds ratio for x. Does it make sense to make an inference about this quantity in this example? Why or why not?

5. Let x take the values -1, $-.5$, 0, $.5$, and 1. At each of these values, generate a sample of 4 values from the *Bernoulli*$(.5)$ distribution. Carry out a logistic regression analysis of this data using the model

$$\ln\left(\frac{p_x}{1-p_x}\right) = \beta_0 + \beta_1 x + \beta_2 x^2$$

Test the null hypothesis $H_0 : \beta_1 = \beta_2 = 0$ versus $H_a : \beta_1 \neq 0$ or $\beta_2 \neq 0$.

Appendix A

Projects

The basic structural component of Minitab is the worksheet. When working on a project, it may make sense to have your data in several worksheets so that similar variables are grouped together. Also, you may wish to save plots associated with the worksheets so that everything can be obtained via a single reference. Worksheets and graphs can be grouped together into *projects*. Projects are given names and are stored in a file with the supplied name and the file extension .mpj.

To open a new project use File ▶ New and choose Minitab Project and click OK. If you want to open a previously saved project, use File ▶ Open Project and choose the relevant project from the list. To save a project, use File ▶ Save Project if the project already has a name (or you wish to use the default of minitab) or File ▶ Save Project As if you wish to give the project a name. Not only are the contents of all worksheets and graphs saved, but the contents of the History folder in the Project window are saved as well and are available when the project is reopened. You can also supply a description of the project using File ▶ Project Description and filling in the dialog box. Note that a description of a worksheet can also be saved using Editor ▶ Worksheet ▶ Description. When you attempt to open a new project or exit Minitab, you will be asked if you wish to save the contents of the current project.

Now suppose that in the project **evans** we have a single worksheet containing 100 numeric values in each of C1 and C2 and have produced a scatterplot of C2 against C1. We open a new worksheet using File ▶ New and choose Minitab Worksheet and click OK. There are now two worksheets associated with the project called Worksheet1 and Worksheet2. Suppose that we also place 100 numeric values in C1 and C2 in Worksheet 2 and again plot C2 against C1. We then have two plots associated with the project **evans** called Worksheet 1: Plot C2*C1 and Worksheet 2: Plot C2*C1. These will all appear as individual windows on your screen, perhaps with some hidden, and any one in particular can be made active by clicking in that window or by clicking on the relevant entry in the list obtained when you use Window. You can also save individual worksheets in the project to files outside the project when a particular worksheet is active using File ▶ Save Current Worksheet As. Similarly, when a graph window is active a graph in the project can be saved to a file outside the project using File ▶ Save Graph As.

With multiple worksheets in a project, it is easy to move data between worksheets using cut, copy, and paste operations. For example, suppose that we want to copy C1 and C2 of Worksheet 1 into C3 and C4 of Worksheet 2. With Worksheet 1 active, highlight the entries in C1 and C2, use Edit ▶ Copy Cells, make Worksheet 2 active, click in the first cell of C3, and use Edit ▶ Paste Cells.

It is possible to see what a project contains without opening it. To do this, use File ▶ Open Project, click on the project to be previewed and click on the Preview button. Similarly, worksheets can be previewed using File ▶ Open Worksheet, clicking on the worksheet to be previewed and clicking on the Preview button.

Appendix B

Mathematical and Statistical Functions in Minitab

B.1 Mathematical Functions

Here is a list and description of the mathematical and statistical functions available in Minitab. All of these functions operate on each element of a column and return a column of the same length. Let $(x_1, \ldots, x_n)$ denote a column of length n. These functions can be applied only to numerical variables.

absolute - computes the absolute value, $(|x_1|, \ldots, |x_n|)$.
antilog - computes the inverse of the base 10 logarithm, $(10^{x_1}, \ldots, 10^{x_n})$.
acos - computes the inverse cosine function, $(\arccos(x_1), \ldots, \arccos(x_n))$.
asin - computes the inverse sine function, $(\arcsin(x_1), \ldots, \arcsin(x_n))$.
atan - computes the inverse tan function, $(\arctan(x_1), \ldots, \arctan(x_n))$.
cos - computes the cosine function when angle is given in radians, $(\cos(x_1), \ldots, \cos(x_n))$.
ceiling - computes the smallest integer bigger than a number, $(\lceil x_1 \rceil, \ldots, \lceil x_n \rceil)$.
degrees - computes the degree measurement of an angle given in radians.
exponentiate - computes the exponential function, $(e^{x_1}, \ldots e^{x_n})$.
floor - computes the greatest integer smaller than a number, $(\lfloor x_1 \rfloor, \ldots, \lfloor x_n \rfloor)$.
gamma - computes the gamma function, $(\Gamma(x_1), \ldots, \Gamma(x_n))$; note that for nonnegative integer x, $\Gamma(x+1) = x!$.
lag - computes the column $(*, x_1, \ldots, x_{n-1})$.
log-gamma - computes the log-gamma function, $(\ln \Gamma(x_1), \ldots, \ln \Gamma(x_n))$; note that for nonnegative integer x, $\ln \Gamma(x+1) = \sum_{i=1}^{x} \ln(i)$.
loge - computes the natural logarithm function, $(\ln(x_1), \ldots, \ln(x_n))$.

193

logten - computes the base 10 logarithm function, $(\log_{10}(x_1), \ldots, \log_{10}(x_n))$.

nscore - computes the normal scores function; see **help**.

parsums - computes the column of partial sums,

$\quad (x_1, x_1 + x_2, \ldots, x_1 + \cdots x_n)$.

parproducts - computes the column of partial products,

$\quad (x_1, x_1 x_2, \ldots, x_1 \cdot \cdots \cdot x_n)$.

radians - computes the radian measurement of an angle given in degrees.

rank - computes the ranks of the column entries, $(r_1, \ldots r_n)$.

round - computes the nearest integer function $i(x)$ with rounding up at .5,

$\quad (i(x_1), \ldots, i(x_n))$; see **help** for more details on this function.

signs - computes the sign function

$$s(x) = \begin{cases} -1 & if \quad x < 0 \\ 0 & if \quad x = 0 \\ 1 & if \quad x > 0 \end{cases}$$

$\quad (s(x_1), \ldots, s(x_n))$.

sin - computes the sine function when the angle is given in radians,

$\quad (\sin(x_1), \ldots, \sin(x_n))$.

sort - computes the column consisting of the sorted (ascending) column entries,

$(x_{(1)}, \ldots, x_{(n)})$.

sqrt - computes the square root function, $(\sqrt{x_1}, \ldots, \sqrt{x_n})$.

tan - computes the tan function when the angle is given in radians,

$\quad (\tan(x_1), \ldots, \tan(x_n))$.

B.2   Column Statistics

Let $(x_1, \ldots, x_n)$ denote a column of length n. Output is written on the screen or in the Session window and can be assigned to a constant. The general syntax for column statistic commands is

column statistic name(E_1)

where the operation is carried out on the entries in column E_1 and output is written to the screen unless it is assigned to a constant using the **let** command.

max - computes the maximum of a column, $x_{(n)}$.

mean - computes the mean of a column, $\bar{x} = (x_1 + \cdots x_n)/n$.

median - computes the median of a column (see Chapter 1).

min - computes the minimum of a column, $x_{(1)}$.

n - computes the number of nonmissing values in the column.

nmiss - computes the number of missing values in the column.

range - computes the difference between the smallest and largest value in a column, $x_{(n)} - x_{(1)}$.

ssq - computes the sum of squares of a column, $x_1^2 + \cdots + x_n^2$.

stdev - computes the standard deviation of a column,

$$s = \sqrt{\frac{1}{n-1}\left[(x_1 - \bar{x})^2 + \cdots + (x_n - \bar{x})^2\right]}.$$

sum - computes the sum of the column entries, $x_1 + \cdots x_n$.

B.3 Row Statistics

Let $(x_1, \ldots, x_n)$ denote a row of length n. The general syntax is

row statistic name $E_1 \ldots E_m\ E_{m+1}$

where the operations are carried out on the rows in columns $E_1, \ldots, E_m$ and the output is placed in column E_{m+1}.

rmax - computes the maximum of a row, $x_{(n)}$.
rmean - computes the mean of a row, $\bar{x} = (x_1 + \cdots x_n)/n$.
rmiss - computes the number of missing values in the row.
rn - computes the number of nonmissing values in the row.
rrange - computes the difference between the smallest and largest value in a row, $x_{(n)} - x_{(1)}$.
rssq - computes the sum of squares of a row, $x_1^2 + \cdots + x_n^2$.
rstdev - computes the standard deviation of a row,

$$s = \sqrt{\frac{1}{n-1}\left[(x_1 - \bar{x})^2 + \cdots + (x_n - \bar{x})^2\right]}.$$

rsum - computes the sum of the row entries, $x_1 + \cdots x_n$.

Appendix C

Macros and Execs

We can store Minitab commands in files that can be called on to execute these commands without having to type them. Also, we can program Minitab to carry out iterative calculations. These aspects involve us in a discussion of *macros* and *execs* in Minitab. We present a very brief overview of this topic and refer the reader to reference [2] in Appendix F for a more extensive discussion. There are two types of macros, *global* and *local*. If you are using Version 12 or later, you need to read only about global and local macros; users of earlier versions need to read only about execs. The capabilities of execs are covered by macros.

Note that because it is possible to write a macro or exec that will loop endlessly, it is important to know how to stop the execution if you feel it is running too long. To do this, simultaneously press the Control and Break keys.

If the macro processor in Minitab finds an error in a macro, this is indicated in the Session window by **ERROR**. If there is an error in a Minitab command in the macro, this is indicated by *ERROR*. Minitab also provides a message that attempts to diagnose what has caused the problem.

C.1 Global Macros

A global macro is a set of commands in a file with the structure

    gmacro
    template
    body
    endmacro

where *template* is a name for the macro, consisting of any characters but starting with a letter, and *body* is a set of Minitab commands, macro statements, or other macro names. In general, it is good form to use the file name for *template* but this is not necessary. If the file has the extension .mac, then only the file name, immediately preceded by %, needs to be used to invoke it. Otherwise, the full file name and extension must be used and again immediately preceded by %. Also, the full path name must be used unless the file is in the default Minitab directory or in the **macros** subdirectory of the default Minitab directory. The statements **gmacro** and **endmacro** must always start and end the file, respectively.

Suppose that we have placed the following statements

```
gmacro
generate
note This macro generates 50 samples of size 20 from the Uniform(0,1).
do k1=1:50
random 20 c1;
uniform 0 1.
let c2(k1)=mean(c1)
enddo
endmacro
```

in a file called **generate.txt**. The macro is invoked via the Minitab command

```
MTB > %generate.txt
```

and causes 50 samples of size 20 to be generated from the uniform distribution on the interval $(0, 1)$, with each sample stored in C1 overwriting the preceding one, and causes the sample mean to be computed for each of these and to be stored in the corresponding element of C2. Finally, a histogram is produced of these 50 means. Clearly, this is a much more powerful method for carrying out simulations than the one we discussed earlier, as that was in essence limited by the size of the worksheet. Note the use of the **do, enddo** statements to perform the calculations iteratively. Also, the **note** command is used to display the text following it, on the same line, in the Session window. Otherwise, there is nothing printed in the Session window beyond the output from any commands that print to this window. If you want the code to be printed in the Session window, place an **echo** command before the code you want printed and a **noecho** command when you want to turn this off.

Macros can be nested; i.e., a macro may have in its body a statement of the form **%file** where **file** contains a macro.

C.1.1   Control Statements

There are a number of statements that allow for control over the order of execution of Minitab commands in a macro.

IF, ELSEIF, ELSE, ENDIF

The **if, elseif, else, endif** command appears in the following structure

```
if expression1
block1
elseif expression2
block2
else
block3
endif
```

where *expression1* and *expression2* are logical expressions and *block1, block2,* and *block3* are blocks of Minitab code. If *expression1* is true, *block1* is executed, if *expression1* is false and *expression2* is true, *block2* is executed; and if both *expression1* and *expression2* are false, *block3* is executed. Note that if one of the expressions is a column of logical values, the expression evaluates as false if all entries are false and as true otherwise. There can be up to 50 **elseif** statements between **if** and **endif.** A logical expression is any expression involving comparison and logical operators that evaluates to true (1) or false (0). For example, the code

```
gmacro
uniform
random 100 c1;
uniform -1 1.
let k1=mean(c1)
if k1<-.5
let k2=0
elseif k1>-.5 and k1<0
let k2=1
elseif k1>0 and k1<.5
let k2=2
else
let k2=3
endif
print k2
endmacro
```

generates a sample of 100 from the uniform distribution on the interval $(-1, 1)$; computes the mean; and outputs 0 if the mean is in $(-1, -.5)$, outputs 1 if the mean is in $(-.5, 0)$, outputs 2 if the mean is in $(0, .5)$, and outputs 3 if it is in $(.5, 1)$.

DO, ENDDO

We saw an example of **do, enddo** in C.1. These statements appear in the following structure

```
do Ki = list
block
enddo
```

where *list* is a list of numbers, perhaps a patterned list such as $-8 : 8/2$, or stored constants. The Minitab code in *block* is executed for each value in the *list* with the constant Ki taking on that value. The numbers in the list can be in increasing or decreasing order.

WHILE, ENDWHILE

The **while, endwhile** statements appear in the following structure

```
while expression
block
endwhile
```

where *expression* is a logical expression and the code in *block* is executed as long as *expression* is true. For example, the code

```
gmacro
stuff
random 2 c1;
uniform 0 1.
let k2=1
let k1=c1(k2)
while k1<.5
let k2=k2+1
if k1<2
let k1=c1(k2)
else
break
endif
endwhile
print k2
endmacro
```

generates a sample of 2 from the Uniform$(0, 1)$, finds the first value in the sample greater than .5, prints its location in the sample, and prints 3 if no such value is found.

NEXT

The **next** command can appear in a **do, enddo** or **while, endwhile** and passes control to the first statement after the **do** or **while,** whichever is relevant, and the loop variable is set to the next value in the list.

BREAK

The **break** command can appear in a **do, enddo** or **while, endwhile** and passes control to the first statement after the **enddo** or **endwhile,** whichever is relevant.

GOTO, MLABEL

The **goto** command allows the macro to skip over a number of statements in the file. This takes the following form

> goto V
> ⋮
> mlabel V

where the goto V statement passes control to the statement following mlabel V and V is a number.

CALL, RETURN

Macros can be invoked from within macros by using statements of the form **%file**. This requires that the macros are in different files. In fact, the macros can be in the same file, all having their own **gmacro** and **endmacro** statements and *templates*. When the file is invoked, the first macro is processed. If the first macro needs to refer to the other macros in the file, this is done via the **call** and **return** commands. For example, suppose that a file contains two macros and the first macro needs to use the second one. This is implemented via the structure

> gmacro
> *template1*
> *body1*
> endmacro
>
> gmacro
> *template2*
> *body2*
> endmacro

where somewhere in *body1* there is the statement

> call *template2*

which transfers control from the first macro to the second macro and somewhere in *body2* there is the statement

> return

which returns control to the first macro.

EXIT

The **exit** command stops the macro. A typical use would be as part of an **if, elseif, else, endif,** where if a certain condition was satisfied no further statements in the macro are executed.

PAUSE, RESUME

The **pause** command returns control to the Session window and session commands can then be invoked. Control is returned to the macro after a **resume** command is issued in the Session window.

C.1.2   Startup Macro

You can place commands that you want to be executed every time you start or restart Minitab in a file called `startup.mac` in the default Minitab directory or the `macro` subdirectory of this directory. For example, you can use the **note** command in such a file to send yourself reminders or the **outfile** command if you always want to record your work in a particular file.

C.1.3   Interactive Macros

A macro can write data to the Session window and accept input from the user. We have already discussed the **note** command, which allows you to write comments to the Session window. The **write** command can be used to write the contents of columns and constants to the Session window. For example, the code

```
gmacro
stuff
random 100 c1;
uniform 0 1.
write "terminal" c1
endmacro
```

generates a sample of 100 from the Uniform(0,1) distribution into C1 and then writes this on the Session window. Of course, we also could have accomplished this using the **print** command but recall that **write** allows for formatted output.

Input can be provided to a macro from the keyboard while the macro is running. This is carried out using the special file name `terminal` with the **read, set,** or **insert** commands. For example, the code

```
note Read 10 observations into C1.
set c1;
file "terminal";
nobs=10.
```

prompts for 10 observations, which are placed into C1. Note the use of the subcommands **file** and **nobs** to the **set** command. You can also use the **read, set,** and **insert** commands in a macro with an **end** statement provided you place the data in the file as well. Also, data can be read in from an external file but the name of the external file must be on the same line as the **file** subcommand and not on the same line as the command as in the session command and, of course, enclosed in single quotes.

The **yesno** command allows you to decide which commands you would like executed perhaps based on what the exec has already computed. For example, the code

note : Would you like to execute the macro random.txt?
yesno k1
if k1=1
%random.txt
endif

asks whether or not you wish to execute the macro in `random.txt`. If you answer `y`, K1 is given the value 1 and the macro `random.txt` is executed; if you answer `n`, K1 is given the value 0 and the macro `random.txt` is not executed.

C.2   Local Macros

Local macros are more sophisticated than global macros. Basically, all the features we have discussed for global macros can also be used in local macros. The major difference is that global macros operate only on the worksheet while local macros create temporary *local worksheets,* which are used for computations without disturbing the *global worksheet.* The contents of local worksheets are not seen in the Session window. Also, local macros can have arguments and subcommands. It is through arguments, such as columns, constants, etc., which are passed to and passed out of the macro, that a local macro operates on the global worksheet. Subcommands to a local macro modify the behavior of the macro. Perhaps local macros are most useful when you want to create a truly new command in Minitab that behaves like the other commands we have been discussing throughout this manual. Because of their considerably more sophisticated nature, we do not discuss local macros any further here and refer the reader to [2] in Appendix F.

C.3   Execs

An exec is a file of Minitab commands that can be executed using the **execute** command. This is convenient if a set of commands is repeatedly used, as then we need only put the set in a file and execute the file. Actually, in Version 12 execs are still available but they have been superseded by global and local macros, which have all the capabilities of execs and more. So if you are using Version 12, you can ignore this section and use macros exclusively.

C.3.1   Creating and Using an Exec

You can use any editor to create a file of Minitab commands. Save the file as ASCII text. If you save the file with the file extension `.mtb`, you can execute the file with the **execute** command by referring only to the file name while otherwise you must provide the file name and extension. If the file is not in the default

Minitab directory, you must provide the full pathname. Alternatively, you can use the **journal** command, described in I.11.8, which places the commands you type in a file with the file extension `.mtj`. If you use this approach, you must refer to the file name and extension.

For example, if the file `random.txt` in the default Minitab directory contains

```
echo
random 100 c1
let k1=mean(c1)
let k2=stdev(c1)
print k1 k2
histogram c1
```

then the command

```
MTB > execute 'random.txt' 3
Executing from file:  random.txt
```

causes the contents of `random.txt` to be executed three times. The **echo** command causes the commands in the file to be printed in the Session window each time they are executed and can be left out if this is not desirable. If you want only some of the commands to be printed, place the **echo** command before those you want printed and the **noecho** command before those you don't want printed. Any comments — lines preceded by # — will not be printed even after an **echo** command. If you want a comment printed then place the comment on a line with the **note** command.

The general syntax of the **execute** command is

execute 'file' V

where V indicates the number of times the commands in `file` are to be executed with V = 1 being the default. If V < 1, the exec is not executed. This can be a useful feature when we want to conditionally execute the exec. For example, V could be the constant K1, which, if it took the value 0 or less, would cause the exec not to be executed.

It is also possible for one exec to call another. In fact, there can be 5 levels of nested execs.

In the window environment, the exec can be executed using File ▶ Other Files ▶ Run an Exec.

C.3.2 The CK Capability for Looping

A form of looping can be carried out using execs. We can denote a generic column by CKi and then use the constant Ki to iterate from column to column. For example, if the worksheet has numeric data in columns C1–C100 and the file `stuff.txt` contains

```
let c101(k1)=mean(ck1)
add k1 1 k1
```

then the commands

```
MTB > let k1=1
MTB > execute 'stuff.txt' 100
```

cause the mean of each of the columns C1–C100 to be computed and to be placed in the corresponding entry of C101. This kind of iteration can also be carried out for matrices (see Appendix D) using the MK capability.

It is possible to use the CK capability to deal with a variable number of columns. For example, suppose that you want to calculate the row means for the columns in a number of worksheets and the number of columns varies from worksheet to worksheet. The exec in the file **rowmeans.txt** with commands

```
let k2=k1+1
rmeans c1-ck1 ck2
```

can do this for each worksheet as in

```
MTB > let k1=30
MTB > execute 'rowmeans.txt'
Executing from file:  rowmeans.txt
```

where we specify that the worksheet has **k1=30** columns and thus the row means are placed in C31.

C.3.3 Interactive Execs

Input can be provided to an exec from the keyboard while the exec is running. This is carried out using the special file name **terminal** with the **read, set,** or **insert** commands. For example, suppose the file **input.txt** contains

```
note Read 10 observations into C1.
set c1;
file "terminal";
nobs=10.
```

When it is executed, we are prompted for 10 observations which are placed into C1.

The **yesno** command allows you to decide which commands you would like executed perhaps based on what the exec has already computed. For example, suppose that the file **adjust.txt** contains the commands

```
let k1=mean(c1)
print k1
note Would you like random.txt to be executed?
yesno k2
execute 'random.txt' k2
```

which calculate the mean of C1, print this out and ask whether or not you wish to execute the exec in `random.txt`. If you answer y, then K2 is given the value 1 and `random.txt` is executed once. If you answer n, then K2 is given the value 0 and `random.txt` is not executed. So, dependent upon what you see for the mean of C1, you can decide whether or not to execute the exec in `random.txt`.

C.3.4 Startup Execs

You can place commands that you want to be executed every time you start or restart Minitab in a file called `startup.mtb` in the default Minitab directory. For example, you can use the **note** command in such a file to send yourself reminders or the **outfile** command if you always want to record your work in a particular file.

Appendix D

Matrix Algebra in Minitab

Some versions of Minitab also have the facility for carrying out matrix algebra. This is useful sometimes as matrices, which can really simplify some complicated algebra and numerical work. In particular, the computations associated with fitting the regression models discussed in sections II.10 and II.11 can be easily handled using matrix algebra. In this appendix we assume that you have been introduced to the basic operations and concepts of matrix algebra.

As an example, consider fitting a quadratic polynomial $\beta_1 + \beta_2 x + \beta_3 x^2$ to n data points $(x_1, y_1), \ldots, (x_n, y_n)$. To do this, we must first create the matrices

$$X = \begin{pmatrix} 1 & x_1 & x_1^2 \\ \vdots & \vdots & \vdots \\ 1 & x_n & x_n^2 \end{pmatrix}$$

and

$$y = \begin{pmatrix} y_1 \\ \vdots \\ y_n \end{pmatrix}$$

The matrix X is called the *design matrix*. In a more advanced statistics course, it is shown that best fitting quadratic (least-squares quadratic) is given by $b_1 + b_2 x + b_3 x^2$, where

$$b = \begin{pmatrix} b_1 \\ b_2 \\ b_3 \end{pmatrix} = \left(X^t X \right)^{-1} X^t y,$$

the vector of predicted values is given by

$$\hat{y} = Xb,$$

the residuals are given by

$$r = y - \hat{y},$$

207

and

$$s^2 = \frac{(y - \hat{y})^t (y - \hat{y})}{n - 3}$$

is the estimate of σ^2.

For the general linear model $E[y] = \beta_1 x_1 + \beta_2 x_2 + \cdots + \beta_k x_k$, where $x_1, \ldots, x_k$ are the explanatory variables and we observe the data $(y_i, x_{1i}, \ldots, x_{ki})$ for $i = 1, \ldots, n$, we present the data in matrix form as

$$X = \begin{pmatrix} x_{11} & \cdots & x_{1k} \\ \vdots & \vdots & \vdots \\ x_{n1} & \cdots & x_{nk} \end{pmatrix}$$

$$y = \begin{pmatrix} y_1 \\ \vdots \\ y_n \end{pmatrix}$$

The best-fitting linear model is $b_1 x_1 + b_2 x_2 + \cdots + b_k x_k$, where

$$b = \left(X^t X\right)^{-1} X^t y$$

$$\hat{y} = Xb$$

$$r = y - \hat{y}$$

and

$$s^2 = \frac{(y - \hat{y})^t (y - \hat{y})}{n - k}.$$

Notice that these formulas are the same for every linear model. Many other useful quantities associated with the linear model can be defined in terms of matrices.

D.1 Creating Matrices

In this section, we illustrate some commands for creating and operating on matrices. We describe the session commands and note the corresponding menu commands.

Matrices in Minitab are denoted by M1, M2, ..., M100. Note that there can be at most 100 matrices. The **name** command can be used to give alternative names to matrices. For example, the command

```
MTB > name m1 'design'
```

assigns the name **design** to the matrix M1 and it can be referred to as such afterward with the name in single quotes.

If we are going to use matrices, the first step is to create them. This can be done in a number of ways. For example, we can use the **read** command as in

```
MTB > read 5 3 m1
DATA> 1 1 1
DATA> 1 2 4
DATA> 1 3 9
DATA> 1 4 16
DATA> 1 5 25
 5 rows read.
MTB > print m1
 Matrix m1
 1 1  1
 1 2  4
 1 3  9
 1 4 16
 1 5 25
```

which creates a 5×3 matrix M1 for the fitting of a quadratic at the x points 1, 2, 3, 4, 5. Note that the dimensions of the matrix accompany the **read** command with the number of rows followed by the number of columns and no **end** statement is required. The corresponding menu command is C̦alc ▶ M̲atrices ▶ R̲ead. Also, matrices can be directly read in from a file using C̦alc ▶ M̲atrices ▶ R̲ead.

Sometimes, you want a matrix with constant entries. The **define** command is available for this. The general syntax of this command is

define V D_1 D_2 E_1

which creates a matrix E_1 with D_1 rows, D_2 columns and every entry is the number V. The corresponding menu command is C̦alc ▶ M̲atrices ▶ Def̲ine Constant.

Often, you want to create a matrix with given entries along its diagonal and 0's in all the off-diagonal elements. The **diagonal** command with syntax

diagonal E_1 E_2

creates a square matrix E_2 with column E_1 in the diagonal and all other entries 0. The matrix E_2 is square with dimension equal to the length of E_1. If instead E_1 is a matrix and E_2 is a column, the diagonal of E_1 is placed into the column E_2. The corresponding menu command is C̦alc ▶ M̲atrices ▶ D̲iagonal.

It is often convenient to copy the content of columns in a worksheet directly into a matrix and vice versa. For example, the commands

```
MTB > set c1
DATA> 5(1)
DATA> end
MTB > set c2
DATA> 1:5
DATA> end
MTB > let c3=c2*c2
MTB > copy c1 c2 c3 m1
```

create the matrix M1, printed above, fairly quickly using shortcut operations associated with the **set** command and basic column operations. For large patterned matrices, this is probably the best way to create the matrix. Also, if the matrix is in an external file we can read the matrix into a set of columns using the **read** command and then use the **copy** command to create the matrix. If M1 is as above the command

    MTB > copy m1 c1-c3

copies the first column of M1 into C1, the second column of M1 into C2, etc. Also, we can create copies of matrices using the **copy** command. For example,

    MTB > copy m1 m2

creates a matrix M2 with the same entries as M1. The corresponding menu command is Calc ▶ Matrices ▶ Copy.

To delete matrices, use the **erase** command. For example,

    MTB > erase m1

deletes the matrix M1.

D.2   Commands for Matrix Operations

There are a variety of commands for performing calculations with matrices. We describe here some session commands for this and give their menu analogs.

add E_1 E_2 E_3 - puts $E_1 + E_2$ into E_3 where E_1, E_2, E_3, are matrices of the same dimension. (Calc ▶ Matrices ▶ Arithmetic)

eigen E_1 E_2 E_3 - puts the eigenvalues of symmetric matrix E_1 into column E_2 and the eigenvectors into matrix E_3. (Calc ▶ Matrices ▶ Eigen Analysis)

invert E_1 E_2 - puts $(E_1)^{-1}$ into E_2 for matrix E_1. (Calc ▶ Matrices ▶ Invert)

multiply E_1 E_2 E_3 - puts $E_1 E_2$ into E_3, where E_1 is a constant and E_2, E_3 are matrices of the same dimension or E_1 is a matrix with the same number of columns as the number of rows in matrix E_2. (Calc ▶ Matrices ▶ Arithmetic)

subtract E_1 E_2 E_3 - puts $E_1 - E_2$ into E_3, where E_1, E_2, E_3 are matrices of the same dimension. (Calc ▶ Matrices ▶ Arithmetic)

transpose E_1 E_2 - puts $(E_1)^t$ into E_2 for matrix E_1. (Calc ▶ Matrices ▶ Transpose)

As an example, suppose we consider fitting the least-squares quadratic when we have observed the data $(1, 7.2365)$, $(2, 17.2625)$, $(3, 33.6455)$, $(4, 55.4614)$, and $(5, 82.2756)$. We construct the X matrix as M1 as indicated above, and the y values are placed in the matrix M2. The commands

```
MTB > transpose m1 m3
MTB > multiply m3 m1 m4
MTB > inverse m4 m4
MTB > multiply m4 m3 m5
MTB > multiply m5 m2 m6
MTB > print m6
 Matrix M6

 2.19780
 2.10946
 2.78638
```

compute the least-squares quadratic as $2.19780 + 2.10946x + 2.78638x^2$. The commands

```
MTB > multiply m1 m6 m7
MTB > subtract m2 m7 m8
MTB > transpose m8 m9
MTB > multiply m9 m8 m10
Answer = 0.1880
MTB > let k1=.1880/2
MTB > print m7 m8 k1
```

store the predicted values in M7, the residuals in M8, and s^2 in K1.

Appendix E

Advanced Statistical Methods in Minitab

While this manual has covered many of the features available in Minitab, there are still a number of other useful techniques that, while not relevant to an elementary course, are extremely useful in a variety of contexts. The material in this manual is good preparation for using the more advanced techniques.

We list here the more advanced methods available in Version 12 and refer the reader to reference [3] in Appendix F for details.

- General linear model and general ANOVA

- Variance components

- MANOVA

- Principal components

- Factor analysis

- Discriminant analysis

- Cluster analysis

- Correspondence analysis

- Time series analysis — trend analysis, decomposition, etc.

- Exploratory data analysis

- Quality-control tools — run chart, Pareto chart, etc.

- Reliability and survival analysis — accelerated life testing, probit analysis, etc.

- Design of experiments — factorial designs, response surface, mixture designs, etc.

Appendix F

References

[1] *Meet Minitab*, Release 13 (2000). Minitab Inc.

[2] *User's Guide 1: Data, Graphics, and Macros,* Release 13 (2000). Minitab Inc.

[3] *User's Guide 2: Data Analysis and Quality Tools* (2000). Minitab Inc.

Index